Barcoding Life

DNA barcoding is revolutionizing the way we identify, catalog, and analyze life on Earth. This method enhances our understanding of species diversity in ecosystems and genetic variation within species, accelerating discoveries and our knowledge of Earth's complex web of life. By providing essential genetic insights, DNA barcoding serves as a key tool for species conservation and management.

Barcoding Life: For Exploring and Protecting Biodiversity gives a detailed exploration of DNA barcoding across all domains of life. Researchers will find that DNA barcoding facilitates rapid species identification, streamlining biodiversity studies and accelerating the discovery of new species. Taxonomists will gain valuable insights into species relationships, enabling more accurate classifications and targeted investigations. Policymakers will learn how DNA barcoding informs conservation priorities, guiding effective strategies and optimizing resource allocation.

With its unique focus on preserving biodiversity at sufficient resolution, this book provides insights and practical applications of DNA barcoding for stakeholders dedicated to safeguarding the natural world. Whether you are a scientist, conservationist, or decision-maker, this book is your essential guide to harnessing the power of DNA barcoding for a sustainable future.

Barcoding Life

For Exploring and Protecting Biodiversity

Edited by Siddra Ijaz, Imran Ul Haq,
and Hayssam Mohamed Ali

CRC Press
Taylor & Francis Group
Boca Raton London New York

CRC Press is an imprint of the
Taylor & Francis Group, an **informa** business

Designed cover image: Shutterstock

First edition published 2026
by CRC Press
2385 NW Executive Center Drive, Suite 320, Boca Raton FL 33431

and by CRC Press
4 Park Square, Milton Park, Abingdon, Oxon, OX14 4RN

CRC Press is an imprint of Taylor & Francis Group, LLC

ISBN: 978-1-032-88744-9 (hbk)
ISBN: 978-1-032-88657-2 (pbk)
ISBN: 978-1-003-53892-9 (ebk)

DOI: 10.1201/9781003538929

Typeset in Times
by Apex CoVantage, LLC

Contents

About the Editors

Prof. Dr. Siddra Ijaz is currently serving as a professor at the Centre of Agricultural Biochemistry and Biotechnology (CABB), University of Agriculture, Faisalabad, Pakistan. She is the medalist in her postgraduate studies. She has made significant research contributions, publishing more than 80 papers with an impact factor above 110, 12 books, and over 55 book chapters on well-recognized and well-renowned publishing platforms, including Springer Nature and Taylor and Francis Group. She has served as the managing editor of the international impact factor journal *Pakistan Journal of Agricultural Sciences*. She has supervised and co-supervised 11 PhD students and over 100 MPhil students. She had been a member of the National Curriculum Revision Committee from 2013 to 2014. She is the head of the 'Molecular Biology of Plant Disease Resistance Laboratory' at CABB, UAF, and the curator of the Fungal Molecular Biology Laboratory Culture Collection (FMB-CC-UAF), an affiliated member of the World Federation for Culture Collections (WFCC). She has expertise in molecular biology, and her research focuses on unlocking the genetics involved in plant disease resistance through *R* gene identification and functional characterization for the exploration of resistance pathways and plant disease management. She has made colossal contributions to fungal taxonomy by reporting novel species of fungal pathogens in plants. She is working with a strong belief in the philosophy that disease resistance gene identification and the exploration of resistance pathways in plants are critical components of ensuring food security, as they play a significant role in mitigating crop losses caused by diseases. She has been awarded more than Rs. 50 million in research grants from different national and international funding agencies.

Prof. Dr. Imran Ul Haq is a professor in the Department of Plant Pathology, University of Agriculture, Faisalabad, Pakistan. He has supervised more than 40 graduate students and established the Fungal Molecular Biology Laboratory Culture Collection (FMB-CC-UAF); he is an affiliated member of the World Federation for Culture Collections (WFCC). He has published more than 60 articles, 11 books, four laboratory manuals, and several book chapters. He has made significant contributions to fungal taxonomy by reporting novel species of fungal pathogens in plants. His research interests are fungal molecular taxonomy and nanotechnology integration with other control strategies for sustainable plant disease management.

Dr. Hayssam Mohamed Ali is working at the Botany and Microbiology Department, College of Science, King Saud University. Dr. Hayssam Mohamed Ali has a 'WORLD RANKING OF TOP 2% SCIENTISTS' according to the Stanford Scientists' Ranking released by Stanford University in 2021, 2022, 2023, 2024, and 2025. He has more than 350 published papers, 20 published books, and he has reviewed more than 1,700 papers in highly impacted journals. He is interested in studying plant physiology, biotic and abiotic stresses, sustainability, environmental stresses, plant biotechnology, crop improvement, plant breeding, forestry, and plant biology.

Contributors

Abdullah
Department of Plant Pathology
University of Agriculture
Faisalabad, Pakistan

Amjad Abbas
Department of Plant Pathology
University of Agriculture
Faisalabad, Pakistan

Jam Nazeer Ahmad
Department of Entomology
University of Agriculture
Faisalabad, Pakistan

Hayssam Mohamed Ali
Department of Botany and Microbiology
College of Science
King Saud University
Riyadh, Saudi Arabia

Muhammad Amjad Ali
Department of Plant Pathology
University of Agriculture
Faisalabad, Pakistan

Muhammad Naveed Anwar
Institute of Microbiology
University of Agriculture
Faisalabad, Pakistan

Muhammad Arshad
Department of Entomology
University of Agriculture
Faisalabad, Pakistan

Usman Arshad
Key Laboratory of Tobacco Pest Monitoring and Integrated Management
Tobacco Research Institute of Chinese Academy of Agricultural Sciences
Qingdao, China

Ayesha Ashfaq
Centre of Agricultural Biochemistry and Biotechnology
University of Agriculture
Faisalabad, Pakistan

Saira Azmat
School of Agriculture and Environment
University of Western Australia
Perth, Australia

Mahrukh Babar
Ibadat International University
Islamabad, Pakistan

Misbah Batool
Department of Physiology
Government College University
Faisalabad, Pakistan

Ceren Cer
Directorate of Plant Protection Research Institute
Republic of Türkiye Ministry of Agriculture and Forestry
Bornova-Izmir, Turkey

Barbaros Çetinel
Directorate of Plant Protection Research Institute
Republic of Türkiye Ministry of Agriculture and Forestry
Bornova-Izmir, Turkey

Abdelfattah A. Dababat
International Maize and Wheat Improvement Center (CIMMYT)
Ankara, Türkiye

Muhammad Ehetisham-Ul-Haq
Oilseeds Research Institute
Ayub Agricultural Research Institute
Faisalabad, Pakistan

Nazia Ehsan
Department of Zoology, Wildlife and Fisheries
University of Agriculture
Faisalabad, Pakistan

Ansha Ghaffar
Department of Physiology
Government College University
Faisalabad, Pakistan

Amer Habib
Department of Plant Pathology
University of Agriculture
Faisalabad, Pakistan

Zakia Habib
Centre of Agricultural Biochemistry and Biotechnology
University of Agriculture
Faisalabad, Pakistan

Faiza Hassan
Institute of Physiology & Pharmacology
University of Agriculture
Faisalabad, Pakistan

Ghulam Hussain
Department of Physiology
Government College University
Faisalabad, Pakistan

Arslan Iftikhar
Department of Physiology
Government College University
Faisalabad, Pakistan

Misbah Ijaz
Department of Clinical Medicine & Surgery
University of Agriculture
Faisalabad, Pakistan

Siddra Ijaz
Centre of Agricultural Biochemistry and Biotechnology
University of Agriculture
Faisalabad, Pakistan

Shawana Iqbal
Department of Physiology
Government College University
Faisalabad, Pakistan

Muhammad Jabran
State Key Laboratory for Biology of Plant Diseases
Institute of Plant Protection
Chinese Academy of Agricultural Sciences
Beijing, China

Muhammad Hassan Jahanzaib
Centre of Agricultural Biochemistry and Biotechnology
University of Agriculture
Faisalabad, Pakistan

Aysha Khalid
Department of Plant Pathology
University of Agriculture
Faisalabad, Pakistan

Nabeeha Aslam Khan
Department of Plant Pathology
University of Agriculture
Faisalabad, Pakistan

Rashad Rasool Khan
Department of Entomology
University of Agriculture
Faisalabad, Pakistan

Cemal Kurt
Department of Field Crops
Faculty of Agriculture
University of Çukurova
Adana, Turkey

Muhammad Tahir Mohyud-din
Department of Clinical Medicine & Surgery
University of Agriculture
Faisalabad, Pakistan

Muhammad Mushahid
Department of Animal Breeding and Genetics
University of Agriculture
Faisalabad, Pakistan

Humaira Muzaffar
Department of Physiology
Government College University
Faisalabad, Pakistan

Ahmad Nisar
Department of Plant Pathology
University of Agriculture
Faisalabad, Pakistan

Muhammad Asif Saddique
University of Agriculture
Faisalabad, Pakistan

Aqsa Saleem
Centre of Agricultural Biochemistry and Biotechnology
University of Agriculture
Faisalabad, Pakistan

Ikhlas Shafique
Department of Sustainable Agriculture and Food Security
Institute of Science
University of Çukurova
Adana, Türkiye

Muhammad Umair Sial
Department of Entomology
University of Agriculture
Faisalabad, Pakistan

Muhammad Huzaifa Tanveer
Department of Plant Pathology
University of Agriculture
Faisalabad, Pakistan

Malaika Tariq
Department of Plant Pathology
University of Agriculture
Faisalabad, Pakistan

Tuaseef Tariq
Department of Plant Pathology
University of Agriculture
Faisalabad, Pakistan

Zain ul Abdin
Department of Entomology
University of Agriculture
Faisalabad, Pakistan

Imran Ul Haq
Department of Plant Pathology
University of Agriculture
Faisalabad, Pakistan

Muhammad Usman
Department of Plant Pathology
University of Agriculture
Faisalabad, Pakistan

Muhammad Waris
Department of Plant Pathology
Baluchistan Agriculture College
Quetta, Pakistan

Muhammad Zulkiffal
Wheat Research Institute, AARI
Faisalabad, Pakistan

1 Use of DNA Barcoding for Protecting and Exploring Plant Biodiversity and Socioeconomic Impact

Siddra Ijaz, Muhammad Asif Saddique, Zakia Habib, and Hayssam Mohamed Ali

1.1 INTRODUCTION

Species identification is the cornerstone of both biodiversity assessment and our basic understanding of the biological realm. However, it is hindered primarily due to a scarcity of professional expertise in classification and the constraints of morphology-based identification methods (Mayden, 2002). At its essence, biodiversity is the intricate mosaic of life, including genetic variety within species and species-to-species changes, all of which contribute to the resilience and adaptation of ecosystems (Salgotra & Chauhan, 2023). There are more than 420,000 plant species, according to statistics from the 2022 International Union for Conservation of Nature (IUCN) Red List. However, only some of them have been categorized using conventional taxonomy techniques based on morphological structures or the phenotypic traits of individual organisms (Letsiou et al., 2024). There are intrinsic shortcomings with the existing taxonomy system, particularly because it is based mostly on morphological parameters, which can be influenced by the environment and present unpredictability and indistinctness in the identification process. Despite cryptic plant species that have the same morphology but belong to distinct evolutionary lineages, standard taxonomy approaches are inadequate for reliably identifying plants (Rouhan & Gaudeul, 2021).

The inherent drawbacks of relying solely on morphological qualities can be overcome by including genetic data to produce a more accurate and reliable plant categorization system (Li et al., 2015). DNA barcoding offers the utmost reliable and quick techniques for ascertaining organisms. This approach is very beneficial since DNA is stable and can be found in all tissues because external or environmental factors do not impact it. DNA barcoding is a novel technique for identifying biological materials using brief DNA sequences from the genomes of either nuclei or other organelles, including chloroplasts and mitochondria (Shneyer & Rodionov, 2019). These barcode

DOI: 10.1201/9781003538929-1

regions are standardized DNA sequences that range in length from four hundred to eight hundred base pairs and are usually exclusive to an organism. These sequences are used to categorize a group of organisms, employing techniques like DNA amplification and sequencing, and evaluation using a reference database with comparable sequences from other species (Ali et al., 2014).

DNA barcoding, via integrating two powerful molecular techniques—PCR amplification and sequencing—suggests a hasty and precise method for identifying diverse groups (Antil et al., 2023). Significantly, DNA barcoding may differentiate between new or unknown species at the species level and cryptic species, which are groups of more than one species based on similar morphological characteristics and are thus indistinguishable using conventional taxonomy (Sheth & Thaker, 2017). In addition to advancing studies in plant evolution, ecology, and conservation, DNA barcoding is a useful tool for protecting endemic and endangered species, especially crucial because of the threats to biodiversity posed by human activities like pollution, deforestation, and resource extraction (Gostel & Kress, 2022). Moreover, it is a potential method for validating the legitimacy and tracking the provenance of food items, such as nutritional supplements and processed dietary products. This method also makes it possible to identify indigenous varieties, improve the nutritional value of crops or their products, and promote locally grown fruits, vegetables, and other health-benefiting plants (Frigerio et al., 2019a).

1.2 INTERNATIONAL BARCODING INITIATIVES

Since DNA barcoding has become prevalent in the past ten years, it does not yet have many automated tools or databases. So, the major barcoding projects established by researchers worldwide aim to develop a universal species identification system based on DNA. Two significant worldwide projects that stand out among them are CBOL and iBOL. The iBOL project was started in 2010 by the University of Guelph's Biodiversity Institute of Ontario in Canada. The iBOL collaboration collaborates with over 20 working groups and constituent nodes from 27 countries. CBOL, established in 2004, is an international project aimed at promoting DNA barcoding as a common method of identifying the many plant and animal species that comprise the biodiversity of the Earth (Pawlowski et al., 2012; Adamowicz, 2015).

DNA barcoding is sometimes not considered convenient, especially when dealing with complicated materials like wild potatoes, despite being widely utilized in many plant identification applications. According to Spooner (2009), the ITS and the plastid markers, including *matK* and *trnH-psbA*, cannot produce markers distinct to a certain species. Since barcoding counts on diverse distinguishable species, which might not be extensively studied across different ecological locations, it cannot adequately distinguish between them. Furthermore, databases might not have enough deposited sequences for precise distinction. Recently, novel techniques using species-specific primers have surfaced, facilitating the differentiation of targeted species. According to Frigerio et al. (2019b), species-specific primers distinguish mustard from wheat and rapeseed. For the analysis of complex plant materials, this invention offers a workable solution to the limitations of traditional barcoding approaches.

1.3 DNA BARCODES

DNA barcoding is fundamentally based on scalability, minimalism, and standardization. A standardized 648-base pair (bp) segment found close to the 5′ ends of COI, a mitochondrial gene, has been designated as the universal barcode that has shown promising methods for animal species identification (Ahmed, 2022). This standard barcode is characterized by a high copy number/cell, making amplification from minute or degraded samples easier, maternal inheritance, no recombination with the paternal genome, and a rapid accumulation of mutations. Unfortunately, because most plants mutate more slowly than animals, COI cannot discern between them, leading to the identification of alternative DNA barcodes for plants (Farha Ahmed, 2022).

The primary principle reinforcing DNA barcoding is based on the concept that during species' evolution, particular DNA sequence lengths within the coding and non-coding regions exhibit high conservation (Costa & Antunes, 2012). The attributes of the sequences found in some areas of nuclear, chloroplast, and cytoplasmic mitochondrial DNA make them suitable candidates for DNA barcoding. These specific sequences' utilization allows for the distinction of species and provides a genetic signature that makes species identification more precise and effective. The design and development of universal primers may facilitate effectual PCR amplification, allowing sequencing and bioinformatics analyses to ascertain each currently recognized species (Antil et al., 2023). Unfortunately, a perfect DNA barcode in the plant kingdom is still unattainable.

1.4 NUCLEAR DNA BARCODES

Plant molecular taxonomy has made extensive use of nuclear ribosomal DNA (rDNA) genes that are intervened by the internal transcribed spacer (ITS), which comprises the ITS1 and ITS2 subregions (Nair, 2019). Over 21,000 plant species have been identified using the ITS as a DNA barcode. ITS stands out as a powerful phylogenetic marker due to its more prevalent distribution and rapid evolutionary rate, which makes it more convenient to compare species that exhibit diversions from each other. Additionally, being a non-coding region, it presents a high variation rate, probably due to reduced functional limitations. The ITS region can be proficiently amplified employing conserved primers, or ITS1 and ITS2 can be coupled alternatively with the 5.8S loci to function as barcodes (Letsiou et al., 2024).

The CBOL Plant-based Working Group proposes using it as a complementary locus for taxonomic categorizing but does not advocate using it as a universal DNA barcode in plants (Plant, 2009; Kowalska et al., 2019). The formation of subordinate structural complications and gene paralogues are significant hindrances that complicate the amplification and sequencing procedures (Hollingsworth et al., 2011).

ITS2 is a potential alternative DNA barcode for taxonomy with an astounding 92.7% accuracy rate in identifying species across over 6500 samples comprising 4800 species (Qin et al., 2017). Its secondary structure also offers valuable details that help determine the species. In addition to COI, ITS2 can be a supplemental locus for identifying animal species (Yao et al., 2010). ITS2 has several benefits, including a high diversity that allows for discrimination even between closely related species, amplifiability, conservation that makes it easier to construct universal primers, and shorter length for easier sequencing. However, despite its potential advantages, it has not been considered universally ideal for plants primarily because of its multiple copies with substantial intraspecific diversity and heterogeneity resulting from concerted evolution (Letsiou et al., 2024).

1.5 DNA BARCODES OF THE CHLOROPLAST GENOME

The chloroplast genome is characterized by a circular DNA molecule that ranges in size from 120 to 220 kb. Until now, one hundred functional genes have been accessible for species identification. Apart from single-locus markers, several researchers propose that DNA barcoding could use the complete plastid genome (Li et al., 2015). Chloroplast-derived DNA barcodes are frequently employed in phylogenetic analyses of plants, as primers are easily designed, the gene order in the organelle's genome is conserved, and many copies per cell make amplification more expedient (Yu et al., 2021). Nonetheless, chloroplast genome genes evolve more slowly, as they exhibit a slow evolutionary rate in contrast to nuclear genome genes. Among the chloroplast DNA barcodes, *matK*, *rbcL*, *trnH-psbA*, *rpoB*, *rpoC1*, *trnL-trnF*, *psbK-psbI*, and *atpF-atpH* have been successfully utilized in plant identification (Ali et al., 2014).

Among the genes in chloroplasts that are rapidly evolving is *matK* (maturase K), which is roughly 1550 base pairs in length and encodes maturase K (*matK*), an enzyme required for intertwining type II introns. It has been successfully employed to identify from the family to the species level, demonstrating minimal transition/transversion rates and evident interspecific divergence (Miao et al., 2019). However, it is challenging to employ a universal DNA barcode because of the high rate of substitution and technological problems, mostly related to the construction of the universal primer sets. Nevertheless, *matK* is a powerful marker for identifying gymnosperms, flowering plants, lycophytes, monilophytes, and lycophytes.

More than 90% of the species in the *Orchidaceae* family and fewer than 49% of the nutmeg species can be discriminated against using *matK* (Kress & Erickson, 2007; Newmaster et al., 2013). Fazekas et al. (2008) only managed a 56% success rate when using the *matK* barcode to identify 92 species from 32 taxa. These findings highlight the possibility that a *matK* barcode by itself might not be adequate as a standard barcode.

rbcL is extensively employed in phylogenetic analyses, and more than 50,000 sequences have been archived in GenBank. It is responsible for encoding the major subunit of the essential photosynthesis-related enzyme RUBISCO. It is considered an impending marker for discrimination at the family and genus levels (Skuza et al., 2019). The complete coding sequence must be sequenced using at least two sets of primers due to its length, which is about 1430 base pairs. Nevertheless, *rbcL* sequences evolve slowly and show the least divergence (Kress et al., 2005). Unfortunately, because of its low discriminatory strength, its applicability for species-level identification is restricted.

TrnH-psbA is a highly variable non-coding plastid locus, and including highly conserved regions makes it possible to build universal primers (Santos, 2018).

This non-coding intergenic region has significant rates of insertion/deletion and the greatest sequence divergence, with much potential for species differentiation. The *trnH-psbA* spacer's complicated molecular history, wide range of lengths, and high proportions of insertion and deletion among higher angiosperm families make the alignment of this gene potentially highly uncertain. In addition, this region contains copies of *rps19* in duplicated loci. The length of the *trnH-psbA* region varies greatly, often surpassing one thousand base pairs in monocots and conifers. However, in some groups, it can be much shorter, less than 300 bp, and in bryophytes, it can be even shorter, less than 100 bp (Li et al., 2015).

A major obstacle to using *trnH-psbA* as a universal barcode is that inversions are frequently seen in certain plant lineages. These inversions may lead to erroneous taxonomic classifications and substantial overestimation of genetic divergence. Furthermore, the length of this region varies between families; in some cases, this region contains copies of *rps19* in addition to a pseudogene positioned between *trnH* and *psbA*. This high-length variation poses an alignment challenge even though high-quality bidirectional sequences can be obtained (Srivastava et al., 2016). In genera such as *Solidago*, it has been noted that diminutive *trnH-psbA* spacer regions cannot have enough variation to distinguish many species (Kress et al., 2005). It was suggested by most of the researchers that to achieve sufficient resolution, *trnH-psbA* could be employed with one or more than one locus; it is still a useful barcode for pteridophytes and flowering plants (Chac & Thinh, 2023).

1.6 OTHER PLASTID DNA BARCODES

Chloroplast loci are the most often used locus in DNA barcoding technology nowadays since they have evolved slower than nuclear loci. Apart from the previously mentioned potential barcodes, there are several additional frequently used barcoding markers from plastid regions, including *rpoB* and *rpoC1*, trnL and trnF, atpF-atpH, and psbK-psbI (Thakur et al., 2019). The plastid genes encode the plastid-encoded RNA polymerase components called rpoB and rpoC1. Although they have been used for family-level identification, they are unsuitable for species discernment in various families of kingdom Plantae because of their lagging pace of evolution (Tadini et al., 2020).

Moreover, the psbK and psbI gene loci, a non-coding intergenic spacer region, encode two polypeptides, K and I, having low molecular weight and are part of photosystem II. Since it is conserved, this intergenic spacer region can be effortlessly amplified, sequenced, and aligned using simple PCR techniques. Although its sequence quality and universality are restricted, it demonstrates a strong discriminating supremacy. The CBOL Group recommends using it as a supplemental locus despite its ability to discriminate due to difficulties in reliably obtaining unblemished bidirectional sequences. However, it is a desirable marker for identifying lycophytes, monilophytes, and bryophytes (Letsiou et al., 2024).

Despite its moderate discriminating ability, atpF-atpH (598 to 613 bp in length) can be a universal DNA barcode for species-based classification. Although atpF-atpH sequences are easily amplified by PCR, aligning them might be difficult. As such, it is more useful as a supplement barcode because it has better precision for particular tasks and taxonomic classifications (Letsiou et al., 2024).

1.7 SINGLE-LOCUS AND MULTI-LOCUS DNA BARCODES

The basic idea behind DNA barcoding is to utilize a DNA sequence, preferably one that is ideally ubiquitous in all species. A criterion for this methodology is establishing a reference database having pertinent sequences from several species (Li et al., 2015). Despite substantial endeavors to discover a universal barcode for plants like CO1 for animals, the efforts have proven challenging due to inadequate discrepancy within individual loci. However, no one locus satisfies the data requirements and locus selection principles set out by CBOL. Thus, utilizing a combination of DNA barcodes has emerged (Plant, 2009; Pawlowski et al., 2012).

Following the CBOL Plant Working Group's guidelines, the barcodes are the most frequently utilized for identifying unknown samples. Because these markers are compatible with universal primer pairs and offer sufficient information on sequence diversity individually and in combination, they are recommended for species differentiation. Researchers have proposed numerous plastid loci combinations, including *rbcL* + *trnH-psbA* (Kress & Erickson, 2007), *rpoC1* + *matK* + *trnH-psbA, or rpoC1* + *rpoB* + *matK* (Chase et al., 2007), and *matK* + *atpF-atpH* + *trnH-psbA* or *matK* + *atpF-atpH* + *psbK-psbI* (Chase et al., 2007). These multi-locus DNA barcode regions unveil greater discriminatory power than single-locus barcodes.

The CBOL Group evaluated the seven previously listed candidate genes and concluded that *matK* + *rbcL* is the universal plant barcode. *MatK* gives greater tenacity but lower universality; however, *rbcL* offers better universality but lower resolution. Together, this combination produces the best species-level discrimination. However, the primary goal of a universal DNA barcode was not met, even though the choice of rbcL + matK led to a somewhat higher identification efficiency when compared to alternative combinations. The rbcL + matK combination primarily does not overcome the matK's poor PCR efficiency (Vijayan & Tsou, 2010; Deka, 2022).

Furthermore, *rbcL* + *matK* usually performs less poorly in plant discrimination than CO1 does in mammals (Ahmed, 2022). Other than these genes, some other potential barcode candidates have been proposed to considerably improve species discrimination, including ITS, ITS2, ycf5, trnL, and different combinations of these genes (Li et al., 2015). Furthermore, the plastid trnH-psbA and the nuclear ITS region have been suggested by Kress et al. (2009) to have potential use as universal plant barcodes. However, this combination can face challenges in species-level identification, especially in low divergence groups, because of the distinct ITS lengths and limited variation offered by trnH-psbA.

According to several scientists, the multi-locus DNA barcodes' inability to improve species discernment was not merely because of the dearth of genetic discrepancies. Rather, it mirrored the inconsistencies between the evolutionary relationship inferred from plastid genes and boundaries in delineating different species (Abdi et al., 2024).

1.8 SUPER BARCODING: A NEW APPROACH FOR DISCRIMINATING PLANTS

Due to the intrinsic precincts of single-locus barcodes, a new technique is required to categorize diligently related plant species. Advances in NGS technology have accelerated the chloroplast (cp) genome use in phylogenetic research. Since the last decade, a plethora of research has utilized the whole cp-genome sequence to address phylogenetic challenges at important evolutionary nodes. It was recently demonstrated that molecular identification, even across closely related species, has been rendered easier by analyzing the chloroplast genome, which produces as much diversity as the COI gene in animals (Dong et al., 2018). The cp genome comprises a conserved region, which ranges between 110 and 160 kb in size and is longer than the standard DNA barcode. It offers extensive variation and makes distinguishing between closely related plants possible. In addition, employing the cp genome as a barcode avoids challenges with PCR efficiency, possible gene deletions, and difficulties retrieving sequences (Zhai et al., 2019).

In phylogenetic, phylogeographic, and population studies of plants, the cp genome is a useful tool that greatly improves resolution at lower taxonomic levels. Its expediency in aiding recovery has led to its suggestion as a species-level DNA barcode. The cp genome is more compact than the nuclear genome and shows lower intraspecific variation and larger interspecific divergence. Superbarcoding outperforms conventional barcoding techniques in identifying gene loss and defining gene order. Superbarcoding relies on improving DNA quality and quantity, lowering sequencing costs, and building a comprehensive chloroplast genome database. Since NGS technology was developed, the sequencing of cp genomes has increased substantially (Kang et al., 2017).

1.9 SPECIFIC DNA BARCODES

The plant barcodes for species-specific discrimination are selected based on the following criteria: assessment of plastid barcode markers suggested by CBOL, commonly employed markers for a specific genus or species identification, screening of hotspot regions via identifying the sequences of oligonucleotide repeats and their association, indels, as well as substitutions distributed in an individual's plastid genome, or the utilization of plastid-based comparative analysis to select a particular genomic region presenting significant divergence at species level. A specific barcode is based on the complete plastid-based comparative analysis that can include either a single-locus barcode such as trnH-psbA and matK or newly identified barcode regions.

The foremost objective of DNA barcoding was identifying a universal barcode region for all plants. Nevertheless, terrestrial plants lack a universal barcode locus, predominantly in the chloroplast, where substitutions are distributed non-randomly and differences specific to lineage evolution may also exist. For this reason, each target group of species' distinct cp-region is used in the particular barcode technique. Any DNA segment with sufficient variability not duplicated inside the cp to prevent analytical difficulties due to paralogy might be utilized as a marker in addition to genes and intergenic spacers. At higher taxonomic levels, including genus and family, single-locus barcodes like *rbcL* can provide a significant resolution; nevertheless, specific barcodes may be required necessarily at this time for species-level identification. A large database of cp-genome sequences (which need not be a replica of the annotated genome of the target taxa) and a distinct database with primers for each species group derived from examining these cp genomes are two requirements for the widespread use of particular barcodes (Letsiou et al., 2024).

The specific barcodes corresponding to the identified species using the primer database can be utilized to differentiate these species. Regarding unidentified species, two procedures will be required. Initially, single-locus barcodes (such as rbcL) are used to classify unknown species at the family/genus levels. The relevant particular DNA barcodes are chosen from cp-genome databases to achieve species-level discrimination. Given the difference in substitution rates, having flexibility in selecting a specific barcode would have significant benefits. In this regard, various barcodes could be identified that are comparable in value to CO1 in animals (Li et al., 2015).

1.10 MINI-BARCODING

Over the past ten years, DNA mini-barcoding has evolved to either extend or circumvent the drawbacks of DNA barcoding. DNA mini-barcoding uses a shorter DNA sequence, typically 200 bp or less than traditional barcodes, which can be efficiently and readily amplified. The main objective of developing a mini-barcode is to identify certain targeted herbal plant species, which are not universal to most species, and to hinder the manipulation of natural herbal commodities. However, mini-barcoding presents difficulties when dealing with intricate herbal combinations that contain ten distinct species. It is crucial to consider the length and location of a particular micro barcode sequence to differentiate between different species, since some DNA sequences may include unhinged mutation sites. Since specific DNA sequences may consist of unstable mutation regions, it is crucial to consider the length and location of a particular mini-barcode sequence to differentiate between different species (Gao et al., 2019; Ahmed, 2022).

1.11 METABARCODING

Metabarcoding uses universal primers to instantaneously generate several DNA barcodes, allowing for the identification of different species in an environmental sample (Ali et al., 2014). The metabarcoding method was assisted by the innovation and advancement of NGS techniques, which permit simultaneous DNA sequence reading without cloning. Metabarcoding intends to identify different or tainted DNA samples, including environmental samples (eDNA), at the family level or even higher (Yu et al., 2011). It contrasts with classical DNA barcoding, which aims to identify integral specimens at the species level. In forensic and ecological investigations, metabarcoding is a flexible and accurate method for examining food matrices to find pollutants or adulterants (Taberlet et al., 2012).

1.12 APPLICATIONS OF DNA BARCODING

In the early years of the 21st century, DNA barcodes, universally reproducible chunks of DNA, were first developed and used for animal identification (Hebert et al., 2004). On the other hand, it took several years for the botanical community to embrace and succeed with a common DNA barcode for plants (Kress, 2017). Based on DNA barcode sequence data, evaluating the degree of genetic variety in plant lineages and the consistency of species designations is possible.

DNA barcoding is a useful approach for biodiversity discovery because it can identify species, particularly cryptic species, that might be unknown to science. DNA barcoding is a tool taxonomy experts use to assess the identification and integrity of botanical products, such as dietary supplements and commercial herbal remedies, and to identify controlled, invasive, and endangered species. Plant DNA barcodes can be developed and applied in two simple processes for identifying purposes. The process involves creating a library of known species' DNA barcodes, and comparing an unknown sample's DNA barcode sequence to the library. Taxonomists must first choose one or several individuals from each species to create reference samples for the DNA barcode library. Using voucher specimens that have been appropriately mounted, labeled, and processed, tissue can be collected directly from living specimens in the field, or removed from specimens already housed in herbaria. These vouchers function as an essential, long-lasting document that links the DNA barcode to a certain plant species.

Community phylogeny and species assembly: Since its inception in 2003, genomics-based research has collaborated with DNA barcoding as a locus-based endeavor (Kress & Erickson, 2008). Both the fields of genomics and DNA barcoding place a strong emphasis on obtaining large-scale genetic data that can solve questions that were previously unsolvable by more data-constrained fields. It is anticipated that approaches and applications from these extremes of gene spectrum will eventually combine sooner rather than later (Coissac et al., 2016). Plant DNA barcodes have been extensively used in the past ten years, particularly in ecology, evolution, and conservation. Ecologists, who examine species interactions and patterns of association, and systematists, who research species identification and evolutionary links, have worked together much more effectively because of DNA barcodes (Baker et al., 2016).

A community phylogeny demonstrates the evolutionary hierarchy among the species that compose a certain ecological community. Plant DNA barcoding is a technique used to understand species' functional characteristics and diversity pools that can impact species composition in an environmental community, as per numerous research studies conducted by ecologist scientists (Swenson, 2012). Furthermore, scientists can construct evolutionary trees that illustrate the species' evolutionary association, depicting how various species have developed in a certain habitat and interrelate with each other. Evolutionary research is essential for comprehending the ecosystem composition and function because of its impact on revealing biodiversity patterns, evolutionary history, and the mechanisms modeling community organization. The composition of a species that develops a species' residence in a certain habitat is substantially affected by various parameters, including composition, species dispersal, and environmental filtration. DNA barcoding provides a precise identification method based on species in a given community, making it an authoritative tool for assessing even minute variations in the species composition and structure that may be underlooked because of outmoded morphological methods (Kress, 2017).

Moreover, DNA barcoding made it possible to identify cryptic plant species that are genetically diverse but similar in terms of morphology, which could fundamentally amend our understanding of community composition and species interaction. Therefore, DNA barcoding offers an unprecedented knowledge of species communications, evolutionary advancements, and biodiversity, establishing an ecosystem in species assembly and investigations related to community phylogeny. Phylomatics, a method for assessing evolutionary trees in plant communities, was a momentous advancement for ecologists (Webb & Donoghue, 2005). The latest studies have been spurred by the community phylogeny for the forests dynamics plot of Barro Colorado island employing DNA barcode regions (Kress et al., 2009), which ultimately accelerated our knowledge regarding species variety and evolutionary relationships (Pei et al., 2011; Kress & Erickson, 2012).

Functional traits and species assembly: Ecologists have long been intrigued by assessing important plant features that enable species to perform in certain environs and amass into communities. An assemblage's degree of trait similarity can be measured to get insight into the characteristics that permit or prevent various species from coexisting. Ecologists have a significant tool for comprehending community-building processes through well-resolved evolutionary histories combined with quantitative data on functional features. DNA barcodes alone cannot reveal any novel information on the significance of functional features in assembling plant species (Swenson, 2012). Nonetheless, there is enough signal in the DNA sequence data to generate phylogenetic conjectures regarding the function of evolutionary signals in species assembly. It was thought that phylogeny's association with traits would enable it to serve as a reliable predictor of trait similarity between species. Unfortunately, there is rarely a direct association between phylogeny and functional features, making it impossible to use phylogenetic signals as a stand-in for ecological similarities (Swenson & Worthy, 2018).

Numerous research studies have integrated functional characteristic data with phylogenetic analyses based on DNA barcoding. Ecologists have investigated the biological processes that influence community assembly in various agroecological zones, from temperate to tropical forest regions (Kress et al., 2009). Baraloto et al. (2021) performed a significant research study in tropical forests, calculating and equating 17 functional attributes in 668 species distributed over nine forest regions in the Northern Amazon. Researchers discovered that co-occurring species showed more functional trait similarities than evolutionary connections. Both parameters significantly influenced the determination of niche overlap. According to their research, environmental filtration is the main factor influencing tree species distribution in these tropical environments.

Similar findings were obtained by Uriarte et al. (2010) in previous research that examined eight attributes in a small cohort of 19 tree species in a Puerto Rican forest plot. Despite a relatively weaker phylogenetic signal, they discovered that at least three features substantially influenced neighborhood structure using a DNA barcode-based community phylogeny. Environmental filtration was determined to be the primary factor shaping this tree community. These studies support Swenson's (2013) conclusion that phylogenetic indicators are not necessarily connected to ecological factors that influence community assembly. However, Swenson also pointed out that both traits have improved the knowledge of community assembly and phylogenetic-based methods, and their potential is still tremendous.

Species delimitation and biodiversity discovery: Morphological characteristics have been used by taxonomists to identify both plants and animals. However, only around 20% of all species on Earth have official names and recognition, despite hundreds of years of labor by taxonomists (Polseela et al., 2016). It still has much work to be done. Determining the borders between species and finding new taxa is made easier with DNA barcoding, a very recent and important technique. Most biodiversity is located in the tropics, which have been instrumental in introducing DNA barcode methods for species discovery, particularly in insect groups. Although DNA barcodes are less effective at discriminating between plants and insects than they are for other groups, botanists have still utilized them in species inventories. Early research mostly examined trees in tropical forests and illustrated the challenges of employing DNA barcodes, particularly given their poor percentage recognition rates of 70% (Dexter et al., 2010).

The investigations also demonstrated notable improvements in distinguishing between juvenile and sterile specimens that lack the conventional morphological characteristics necessary for diagnosis. Their investigation showed that DNA barcoding is useful for quickly evaluating biodiversity and identifying cryptic tree populations. Comparable research conducted in a rainforest plot in central Africa revealed poorer species-level identification success (71–88%) but good genus-level resolution (95–100%), especially in species-rich clades.

Plant taxonomists trying to use DNA barcodes in highly diverse tropical regions encounter several challenges, chief among them being that many species are relatively unknown and lack Latin binomials. Ecologists have built up forest inventory plots to examine the dynamics of forests throughout time along gradients in height, latitude, or habitat. These plots are full of 'morphotypes' that do not yet have scientific names. Building DNA barcode libraries of these taxa can substantially improve the tracking of these morphotypes within plots and when comparing them to recognized species, which can sometimes be challenging and error-prone (Kress, 2017).

DNA barcode forensics: Conserving biodiversity, which is seriously threatened by significant habitat loss and environmental change by human activity, is one of the biggest issues confronting biologists today. As previously said concerning herbal supplements and medications, a major issue in evaluating species variety and identifying items sold in local markets is the absence of standardized taxonomy.

The use of DNA barcoding in conservation is still evolving. However, studies show that standardized and comparable genetic data, like DNA barcode sequences, are effective tools for basic research and conservation, improving conservation tracking, priority evaluations, and biodiversity studies over large geographic areas, including indigenous lands, diverse landscapes, and endangered ecosystems (Shapcott et al., 2015; Erickson et al., 2017).

Future of DNA barcoding in plants: DNA barcoding has been used extensively in basic and applied plant research since it was first introduced more than ten years ago. It hasn't, however, been widely accepted by plant systematists as the main method of species identification. It is mostly because no single genetic signature can accurately classify species across most taxonomic categories. To tackle important scientific issues, ecologists, who usually deal with various plant lineages, have been more open to investigating novel DNA barcoding applications. Plant DNA barcoding is anticipated to progress in two main areas to serve the botanical community better: creating novel genetic markers while using cutting-edge sequencing technology, and growing an extensive worldwide DNA barcode library for greater accessibility (Mahadani et al., 2022).

Over the next ten years, one of the main challenges will be creating a worldwide library of plant DNA barcodes. Trees tagged for individual tracking, herbarium collections, and forest monitoring plots provide useful resources for building and growing this plant DNA barcode library, laying the groundwork for further study and biodiversity monitoring. Furthermore, individual taxonomists create DNA barcodes for specific plant groups in their fundamental molecular phylogenetic research. These DNA barcode markers help clarify evolutionary relationships and serve as tests for sequencing success with standard markers (Wang et al., 2017). Recently, much work has been done to create generic DNA barcodes for whole local floras. The collection designed to identify Canada's vascular plants is among the most outstanding, and 96% of the 5108 species in Canada have barcode sequence data that Braukmann and associates have successfully created (Braukmann et al., 2017).

The effectiveness of covering the species pool for the three markers (rbcL, matK, and ITS2) varied. They found that in subregions with the highest floristic variety, these markers performed the best at discriminating between species and genera. The challenge of building a DNA barcode library is locating financial sources, often provided by government funding agencies to pay for the sequencing and laboratory expenses. Although creating a global library of DNA barcodes for every plant species will take some time, fundamental and applied research will benefit immensely from it.

Novel DNA markers and sequencing technologies: The emergence of studies applying plant DNA barcodes to questions of taxonomy, evolution, and ecology sparked speculation and predictions about the technology almost simultaneously. It was determined that effective database architecture, search techniques, and cutting-edge sequencing technology were necessary for identifying species (Taberlet et al., 2012). For precise species identification in these situations, relatively brief and precise genetic markers are frequently required. These markers may be 'mini barcodes,' focusing on a smaller portion of the conventional barcode sequences rather than the regular DNA barcodes (Hajibabaei & McKenna, 2012).

Due to technological breakthroughs, the discipline of metabarcoding is expanding quickly. Although it remains challenging, developing new bioinformatics methods to translate DNA sequences from samples into lists of recognized species is a future goal. According to some research, genome skimming and low-coverage shotgun sequencing of nuclear and plastid regions could link whole-genome sequences and standard DNA barcoding, providing the most accurate identification of a species. In addition to eliminating the requirement for PCR, this 'mega-barcoding' approach will yield more genetic information that can be used for purposes other than species identification, such as phylogenetic resolution.

1.13 SOCIOECONOMIC IMPACT OF DNA BARCODING IN PLANTS

In plants, DNA barcoding has a profound socioeconomic impact across various sectors, including agriculture, biodiversity conservation, trade, and herbal medicine. In human history, it is estimated that around 7,000 plant species, along with their various cultivars and varieties, have been cultivated for human sustenance. Nevertheless, only a small percentage of these (such as wheat, rice, and maize) supply the majority of humans' dietary energy demand worldwide (Campanaro et al., 2019). Climate change, pesticide resistance, and the spread of pests are significant issues in cultivating major staples, ultimately contributing to global food scarcity, which has substantial social and economic implications. DNA barcoding offers a more accurate and precise approach for identifying plant species, particularly cryptic species, whose identification can result in significant financial losses and even health risks. Exploiting biodiversity, including underutilized and neglected crops, is one potential way to achieve economic development. Accurately identifying species and their variants is essential for authenticating the raw materials, such as seeds and associated processed products, distributed locally, and for acquiring data on their origins. Moreover, DNA barcoding ensures cultivar authenticity and seed purity, which is crucial to the seed industry, as it certifies high-quality plant materials and prevents the introduction of counterfeit germplasm.

Additionally, the role of DNA barcoding is also remarkable in protecting economically important and endangered plant species. It facilitates the implementation of CITES legislation, discourages illicit logging, and assists in identifying protected species in timber trading and other plant product trades. In the food and spice industries, DNA barcoding ensures quality control and protects against economic deception by validating high-value items, such as turmeric, saffron, and cinnamon. By satisfying global safety and labeling regulations, this boosts export opportunities and consumer confidence (Thomas and Sajesh, 2026). Moreover, the international horticulture sector is one of the fastest-growing industries, and its global industry sales have surpassed $60 billion, according to an economic survey by the US Bureau. Several counterfeit decorative and fruit plants have been sold without anyone acquiring intellectual property rights or complying with plant variety protection rights, which has undeniably undermined the global horticulture market. As a result, plant breeders, including those at public research institutes and businesses of various scales, may suffer significant financial losses in horticulture crops. Authorized producers pay most of their revenues and license dues; however, illicit traders disregard these expenditures, which has a detrimental impact on both society and international trade. Hence, DNA barcoding applications provide an authoritative tool for identifying valuable plant species, benefiting the world's economy (Elansary et al., 2017).

Overall, DNA barcoding fosters opportunities in molecular diagnostics, genomics, conservation, and biotechnology that collectively contribute to the expansion of the bioeconomy. Through the development of bioinformatics platforms, educational enterprises, and molecular identification services, it encourages capacity-building and innovation in developing countries. By integrating DNA barcoding into extension services, it empowers farmers in rural areas by combining contemporary scientific approaches with traditional plant data. DNA barcoding promotes a more equitable and sustainable society, with a beneficial impact on society, the environment, the economy, and human well-being, by addressing issues related to product authenticity, illegal trade, and sustainable resource management.

REFERENCES

Abdi, G., S. Singh, S. Selvakumar, S.K. Dhar, G. Mudgal, P. Swaminathan, M. Jain and M.A. Tarighat. 2024. DNA barcoding and its applications. In *Advances in genomics: Methods and applications*. Springer. pp. 91–117.

Adamowicz, S.J. 2015. International barcode of life: Evolution of a global research community. *Genome* 58:151–162.

Ahmed, S.S. 2022. DNA barcoding in plants and animals: A critical review. pp. 1–28.

Ali, M.A., G. Gyulai, N. Hidvegi, B. Kerti, F.M.A. Al Hemaid, A.K. Pandey and J. Lee. 2014. The changing epitome of species identification–DNA barcoding. *Saudi J. Biol. Sci.* 21:204–231.

Antil, S., J.S. Abraham, S. Sripoorna, S. Maurya, J. Dagar, S. Makhija, P. Bhagat, R. Gupta, U. Sood and R. Lal. 2023. DNA barcoding, an effective tool for species identification: A review. *Mol. Biol. Rep.* 50:761–775.

Baker, C.C.M., L.S. Bittleston, J.G. Sanders and N.E. Pierce. 2016. Dissecting host-associated communities with DNA barcodes. *Philos. Trans. R. Soc. B Biol. Sci.* 371:20150328.

Baraloto, C., J. Vleminckx, J. Engel, P. Petronelli, N. Dávila, M. RÍos, E.H. Valderrama Sandoval, I. Mesones, J.E. Guevara Andino and C. Fortunel. 2021. Biogeographic history and habitat specialization shape floristic and phylogenetic composition across Amazonian forests. *Ecol. Monogr.* 91:e01473.

Braukmann, T.W.A., M.L. Kuzmina, J. Sills, E. V Zakharov and P.D.N. Hebert. 2017. Testing the efficacy of DNA barcodes for identifying the vascular plants of Canada. *PLoS One* 12:e0169515.

Campanaro, A., N. Tommasi, L. Guzzetti, A. Galimberti, I. Bruni and M. Labra. 2019. DNA barcoding to promote social awareness and identity of neglected, underutilized plant species having valuable nutritional properties. *Food Res. Int.* 115:1–9.

Chac, L.D. and B.B. Thinh. 2023. Species Identification through DNA Barcoding and its applications: A review. *Biol. Bull.* 50:1143–1156.

Chase, M.W., R.S. Cowan, P.M. Hollingsworth, C. Van Den Berg, S. Madriñán, G. Petersen, O. Seberg, T. Jørgsensen, K.M. Cameron and M. Carine. 2007. A proposal for a standardised protocol to barcode all land plants. *Taxon* 56:295–299.

Coissac, E., P.M. Hollingsworth, S. Lavergne and P. Taberlet. 2016. From barcodes to genomes: Extending the concept of DNA barcoding. pp. 1423–1428.

Costa, F.O. and P.M. Antunes. 2012. The contribution of the barcode of life initiative to the discovery and monitoring of biodiversity. In *Natural resources, sustainability and humanity: A comprehensive view*. pp. 37–68. Dordrecht: Springer Netherlands.

Deka, P.C. 2022. Plant DNA barcoding. In *Research anthology on ecosystem conservation and preserving biodiversity*. IGI Global. pp. 493–515.

Dexter, K.G., T.D. Pennington and C.W. Cunningham. 2010. Using DNA to assess errors in tropical tree identifications: How often are ecologists wrong and when does it matter? *Ecol. Monogr.* 80:267–286.

Dong, W., C. Xu, P. Wu, T. Cheng, J. Yu, S. Zhou and D.-Y. Hong. 2018. Resolving the systematic positions of enigmatic taxa: Manipulating the chloroplast genome data of Saxifragales. *Mol. Phylogenet. Evol.* 126:321–330.

Elansary, H.O, M. Ashfaq, H.M. Ali and K. Yessoufou. 2017. The first initiative of DNA barcoding of ornamental plants from Egypt and potential applications in horticulture industry. *PLoS One* 12(2):e0172170.

Erickson, D.L., E. Reed, P. Ramachandran, N.A. Bourg, W.J. McShea and A. Ottesen. 2017. Interactions between white-tailed deer and invasive plants in North American forests: reconstructing a herbivore's diet using a novel rbcL DNA mini-barcode for plants. *AoB Plants* 9:1–18.

Farha Ahmed, M.K.Z. 2022. A critical review on the challenges and advances in dna barcoding for plant identification. *Curr. Trends Pharm. Res.* 9.

Fazekas, A.J., K.S. Burgess, P.R. Kesanakurti, S.W. Graham, S.G. Newmaster, B.C. Husband, D.M. Percy, M. Hajibabaei and S.C.H. Barrett. 2008. Multiple multilocus DNA barcodes from the plastid genome discriminate plant species equally well. *PLoS One* 3:e2802.

Frigerio, J., T. Gorini, A. Galimberti, I. Bruni, N. Tommasi, V. Mezzasalma and M. Labra. 2019a. DNA barcoding to trace Medicinal and Aromatic Plants from the field to the food supplement. *J. Appl. Bot. Food Qual.* 92:33–38.

Frigerio, J., R. Pellesi, V. Mezzasalma, F. De Mattia, A. Galimberti, F. Lambertini, M. Suman, S. Zanardi, A. Leporati and M. Labra. 2019b. Development of a DNA barcoding-like approach to detect mustard allergens in wheat flours. *Genes (Basel).* 10:234.

Gao, Z., Y. Liu, X. Wang, X. Wei and J. Han. 2019. DNA mini-barcoding: A derived barcoding method for herbal molecular identification. *Front. Plant Sci.* 10:987.

Gostel, M.R. and W.J. Kress. 2022. The expanding role of DNA barcodes: Indispensable tools for ecology, evolution, and conservation. *Diversity* 14:213.

Hajibabaei, M. and C. McKenna. 2012. DNA mini-barcodes. In *DNA barcodes methods Protoc.* 339–353. Totowa, NJ: Humana Press.

Hebert, P.D.N., E.H. Penton, J.M. Burns, D.H. Janzen and W. Hallwachs. 2004. Ten species in one: DNA barcoding reveals cryptic species in the neotropical skipper butterfly Astraptes fulgerator. *Proc. Natl. Acad. Sci.* 101:14812–14817.

Hollingsworth, P.M., S.W. Graham and D.P. Little. 2011. Choosing and using a plant DNA barcode. *PLoS One* 6:e19254.

Kang, Y., Z. Deng, R. Zang and W. Long. 2017. DNA barcoding analysis and phylogenetic relationships of tree species in tropical cloud forests. *Sci. Rep.* 7:12564.

Kowalska, Z., F. Pniewski and A. Latała. 2019. DNA barcoding–A new device in phycologist's toolbox. *Ecohydrol. Hydrobiol.* 19:417–427.

Kress, W.J. 2017. Plant DNA barcodes: Applications today and in the future. *J. Syst. Evol.* 55:291–307.

Kress, W.J. and D.L. Erickson. 2007. A two-locus global DNA barcode for land plants: The coding rbcL gene complements the non-coding trnH-psbA spacer region. *PLoS One* 2:e508.

Kress, W.J. and D.L. Erickson. 2008. DNA barcodes: Genes, genomics, and bioinformatics. *Proc. Natl. Acad. Sci.* 105:2761–2762.

Kress, W.J. and D.L. Erickson. 2012. *DNA barcodes: Methods and protocols*. Springer.

Kress, W.J., D.L. Erickson, F.A. Jones, N.G. Swenson, R. Perez, O. Sanjur and E. Bermingham. 2009. Plant DNA barcodes and a community phylogeny of a tropical forest dynamics plot in Panama. *Proc. Natl. Acad. Sci.* 106:18621–18626.

Kress, W.J., K.J. Wurdack, E.A. Zimmer, L.A. Weigt and D.H. Janzen. 2005. Use of DNA barcodes to identify flowering plants. *Proc. Natl. Acad. Sci.* 102:8369–8374.

Letsiou, S., P. Madesis, E. Vasdekis, C. Montemurro, M.E. Grigoriou, G. Skavdis, V. Moussis, A.E. Koutelidakis and A.G. Tzakos. 2024. DNA barcoding as a plant identification method. *Appl. Sci.* 14:1415.

Li, X., Y. Yang, R.J. Henry, M. Rossetto, Y. Wang and S. Chen. 2015. Plant DNA barcoding: From gene to genome. *Biol. Rev.* 90:157–166.

Mahadani, P., M. Dasgupta, J. Vijayan, C.S. Kar and S. Ray. 2022. DNA barcoding in plants: Past, present, and future. In *Plant genomics for sustainable agriculture*. Springer. pp. 331–350.

Mayden, R.L. 2002. On biological species, species concepts and individuation in the natural world. *Fish Fish.* 3:171–196.

Miao, L.I.U., L.I. Xi-Wen, L. Bao-Sheng, L.U.O. Lu and R.E.N. Yue-Ying. 2019. Species identification of poisonous medicinal plant using DNA barcoding. *Chin. J. Nat. Med.* 17:585–590.

Nair, K.N. 2019. Methods and approaches in plant molecular systematics. Plant taxonomy and biosystematics. 383–416.

Newmaster, S.G., M. Grguric, D. Shanmughanandhan, S. Ramalingam and S. Ragupathy. 2013. DNA barcoding detects contamination and substitution in North American herbal products. *BMC Med.* 11:1–13.

Pawlowski, J., S. Audic, S. Adl, D. Bass, L. Belbahri, C. Berney, S.S. Bowser, I. Cepicka, J. Decelle and M. Dunthorn. 2012. CBOL protist working group: Barcoding eukaryotic richness beyond the animal, plant, and fungal kingdoms. *PLoS Biol.* 10:e1001419.

Pei, N., J. Zhang, X. Mi and X. Ge. 2011. Plant DNA barcodes promote the development of phylogenetic commu-nity ecology. *Biodivers. Sci.* 19:284.

Plant, C. 2009. Group W. A DNA barcode for land plants. *Proc. Natl. Acad. Sci. USA* 106:7.

Polseela, R., N. Jaturas, A. Thanwisai, K.-W. Sing and J.-J. Wilson. 2016. Towards monitoring the sandflies (Diptera: Psychodidae) of Thailand: DNA barcoding the sandflies of Wihan Cave, Uttaradit. *Mitochondrial DNA Part A* 27:3795–3801.

Qin, Y., M. Li, Y. Cao, Y. Gao and W. Zhang. 2017. Molecular thresholds of ITS2 and their implications for molecular evolution and species identification in seed plants. *Sci. Rep.* 7:17316.

Rouhan, G. and M. Gaudeul. 2021. Plant taxonomy: A historical perspective, current challenges, and perspectives. *Mol. Plant Taxon. Methods Protoc.* 1–38.

Salgotra, R.K. and B.S. Chauhan. 2023. Genetic diversity, conservation, and utilization of plant genetic resources. *Genes (Basel).* 14:174.

Santos, C.G. 2018. Development of new tools for the identification of plants using chloroplast DNA sequences (Doctoral dissertation). University of Porto, Porto, Portugal.

Shapcott, A., P.I. Forster, G.P. Guymer, W.J.F. McDonald, D.P. Faith, D. Erickson and W.J. Kress. 2015. Mapping biodiversity and setting conservation priorities for SE Queensland's rainforests using DNA barcoding. *PLoS One* 10:e0122164.

Sheth, B.P. and V.S. Thaker. 2017. DNA barcoding and traditional taxonomy: An integrated approach for biodiversity conservation. *Genome* 60:618–628.

Shneyer, V.S. and A. V Rodionov. 2019. Plant DNA barcodes. *Biol. Bull. Rev.* 9:295–300.

Skuza, L., I. Szućko, E. Filip and A. Adamczyk. 2019. DNA barcoding in selected species and subspecies of Rye (Secale) using three chloroplast loci (matK, rbcL, trnH-psbA). *Not. Bot. Horti Agrobot. Cluj-Napoca* 47:54–62.

Spooner, D.M. 2009. DNA barcoding will frequently fail in complicated groups: An example in wild potatoes. *Am. J. Bot.* 96:1177–1189.

Srivastava, S., Sanchita, M. Bhargava and A. Sharma. 2016. DNA barcoding of medicinal plants. In *Medicinal plants – recent advances in research and development*. pp. 97–127. Singapore: Springer Singapore.

Swenson, N.G. 2012. Phylogenetic analyses of ecological communities using DNA barcode data. In *DNA barcodes methods protocols*. pp. 409–419. Totowa, NJ: Humana Press.

Swenson, N.G. 2013. The assembly of tropical tree communities–the advances and shortcomings of phylogenetic and functional trait analyses. *Ecography (Cop.).* 36:264–276.

Swenson, N.G. and S.J. Worthy. 2018. Phylogenetic resolution and metrics of biodiversity and signal in conservation. In *Phylogenetic diversity: Applications and challenges in biodiversity science*. pp. 93–110. Cham: Springer International Publishing.

Taberlet, P., S.M. Prud'Homme, E. Campione, J. Roy, C. Miquel, W. Shehzad, L. Gielly, D. Rioux, P. Choler and J. Clément. 2012. Soil sampling and isolation of extracellular DNA from large amount of starting material suitable for metabarcoding studies. *Mol. Ecol.* 21:1816–1820.

Tadini, L., N. Jeran, C. Peracchio, S. Masiero, M. Colombo and P. Pesaresi. 2020. The plastid transcription machinery and its coordination with the expression of nuclear genome: Plastid-encoded polymerase, nuclear-encoded polymerase and the genomes uncoupled 1-mediated retrograde communication. *Philos. Trans. R. Soc. B* 375:20190399.

Thakur, V.V., S. Tiwari, N. Tripathi and G. Tiwari. 2019. Molecular identification of medicinal plants with amplicon length polymorphism using universal DNA barcodes of the atpF–atpH, trnL and trnH–psbA regions. *3 Biotech* 9:188.

Thomas, L. and V.K. Sajesh. 2026. Consumer awareness and market demands in global spice value chains. In *Spices production to products*. CRC Press. pp. 343–361.

Uriarte, M., N.G. Swenson, R.L. Chazdon, L.S. Comita, W. John Kress, D. Erickson, J. Forero-Montaña, J.K. Zimmerman and J. Thompson. 2010. Trait similarity, shared ancestry and the structure of neighbourhood interactions in a subtropical wet forest: Implications for community assembly. *Ecol. Lett.* 13:1503–1514.

Vijayan, K. and C.H. Tsou. 2010. DNA barcoding in plants: Taxonomy in a new perspective. *Curr. Sci.* 1530–1541.

Wang, R., J. Chen, J.A. Anderson, J. Zhang, W. Zhao, J. Wheeler, N. Klassen, D.R. See and Y. Dong. 2017. Genome-wide association mapping of Fusarium head blight resistance in spring wheat lines developed in the Pacific Northwest and CIMMYT. *Phytopathology* 107:1486–1495.

Webb, C.O. and M.J. Donoghue. 2005. Phylomatic: Tree assembly for applied phylogenetics. *Mol. Ecol. Notes* 5:181–183.

Yao, H., J. Song, C. Liu, K. Luo, J. Han, Y. Li, X. Pang, H. Xu, Y. Zhu and P. Xiao. 2010. Use of ITS2 region as the universal DNA barcode for plants and animals. *PLoS One* 5:e13102.

Yu, J., X.I. Wu, C. Liu, S. Newmaster, S. Ragupathy and W.J. Kress. 2021. Progress in the use of DNA barcodes in the identification and classification of medicinal plants. *Ecotoxicol. Environ. Saf.* 208:111691.

Yu, J., J. Xue and S. Zhou. 2011. New universal matK primers for DNA barcoding angiosperms. *J. Syst. Evol.* 49:176–181.

Zhai, W., X. Duan, R. Zhang, C. Guo, L. Li, G. Xu, H. Shan, H. Kong and Y. Ren. 2019. Chloroplast genomic data provide new and robust insights into the phylogeny and evolution of the Ranunculaceae. *Mol. Phylogenet. Evol.* 135:12–21.

2 Fungal DNA Barcoding for Species Identification

Barbaros Çetinel, Ceren Cer, Aqsa Saleem, Imran Ul Haq, and Siddra Ijaz

2.1 INTRODUCTION

Fungi belong to the second most diverse eukaryotic kingdom, characterized by heterotrophic traits and a cell wall composed of chitin. Fungi are the major decomposers of the ecosystem, converting dead inorganic matter into useful organic components. Therefore, an accurate and rapid identification of fungal species is important in many fields in the prospectus of biodiversity and ecology. To identify cryptic species of fungi for biodiversity studies, the barcode sequence is a helpful tool.

There were almost 16 presentations given at the 6th International Barcode Conference in Canada in 2015 on various groups and classes of fungi, including edible mushrooms and ubiquitous fungal pathogens (Adamowicz, 2015). Fungi is considered simple in comparison to plants and animals. But this is not true, as fungi show a great deal of diversity in morphology, mode of nutrition, and reproduction (Alexopoulos et al., 1996; Ainsworth et al., 2008). Fungi, morphologically, are seen in many forms, ranging from microscopic unicellular yeasts to filamentous molds and macroscopic mushrooms. Ecologically, a fungus is found in every environment, from hot springs to cold, freezing environments, due to its thick-walled chitin covering. Fungi adapted to every niche from the depth of the ocean to the earth's substratum, from solid rocks (in the form of lichen association) to the stratosphere, from the earth's equator to Antarctic and Atlantic regions, and from deserts to rivers and lakes (Kurtzman et al., 2011).

Fungi multiply both sexually and asexually with a range of reproductive structures, such as conidiophore to notorious club-shaped fungi, many times reproduced sexually or asexually in the form of spores (that can tolerate harsh environmental stress conditions) with the ability to spread or expand over a wide area (Ainsworth et al., 2008; Xu, 2005).

The exponential growth of fungal DNA barcoding research has been greatly facilitated by the widespread availability of cost-effective DNA sequencing technologies. The entire ITS region spans approximately 600 base pairs and comprises two hypervariable sections, ITS1 and ITS2, which are separated by highly conserved 5.8S rRNA, providing a valuable marker for fungal identification and family tree investigation (White et al., 1990). The presence of highly conserved sequences in the 18S, 5.8S, and 28S rRNA genes facilitates the development of universal primer, allowing for the efficient amplification of the ITS1, ITS2, and entire ITS region across a broad range of fungal taxa.

 DOI: 10.1201/9781003538929-2

The use of ITS as a barcode is further bolstered by the fact that fungal genomes often harbor multiple ribosomal RNA gene clusters. This redundancy enables the successful amplification of the ITS gene, even when only a trace amount of biological material is available, rendering it an exceptionally powerful tool for fungal identification. Despite the advances in fungal research, the vast majority of fungal diversity remains unexplored. With estimates suggesting that only around 1% (100000) of the approximately 6–7 million of this species are known, there is still much to be uncovered in the fascinating world of fungi (Taylor et al., 2014). It means having only one DNA sequence per species makes it difficult to understand the genetic diversity within that species and how it relates to other closely related species.

Traditional methods of identifying fungi don't always align with their DNA relationships. Different-looking fungi can have identical DNA sequences. Some confusion about fungal identities arose from differences in appearance at various growth stages rather than from actual differences between species. On the other hand, DNA analysis often reveals that a single fungus species, identified by the traditional method, actually consists of multiple hidden (cryptic) species. As a result, much of the unknown fungal diversity will likely be discovered and characterized through DNA sequencing alone, without the need for live cultures or physical specimens. DNA-based identifications and recognition of fungal species offer a powerful approach to exploring and understanding the vast fungal diversity.

2.1.1 Species: From Concepts to Identification

While often used interchangeably, the processes of defining species conceptualization, delimitation, recognition, identification, and verification are distinct, each requiring unique methodologies. However, these stages are interconnected. For fungi, in particular, accurate identification hinges on first establishing a well-defined species concept. This conceptual foundation enables systematic delimitation and recognition of species based on agreed diagnostic traits, which are essential for developing reliable identification and verification tools.

2.1.2 Species Concepts

Species concepts provide the theoretical basis for naming and categorizing biodiversity across all life forms and are central to constructing identification frameworks. In fungi, for example, sexual and asexual forms were traditionally considered separate species under dual nomenclature. However, with advances in DNA technology, these morphs can now be genetically matched, making the dual nomenclature system outdated and leading to the 'one fungus, one name' paradigm (Hawksworth, 2011; Taylor, 2011; Wingfield et al., 2012; Geiser et al., 2013).

Three main criteria often underpin species delimitation: genealogical coherence (often in the form of monophyly), reproductive isolation, and phenotypic distinctiveness, which can include ecological specializations (Eyualem & Blaxter, 2003). Genealogical species focus on lineage separation, while morphological or phenotypic species are based on visible traits and behavior. Biological and recognition species models emphasize mating compatibility and reproductive isolation

(Plitzner et al., 2019). Some fungal groups include unique types, such as asexual lineages or hybrids. Integrative models, such as the cohesion or evolutionary species concepts, blend genetic, reproductive, and phenotypic criteria. Other concepts, such as the evolutionarily significant unit and the least inclusive taxonomic unit, seek fine resolution in species classification (Moritz, 1994; Wilkins, 2018).

Due to these differing conceptual approaches, the same group of organisms might be categorized differently depending on which model is used. This lack of uniformity can pose challenges for users of identification tools built on conflicting classification modules (Agapow et al., 2004; Taylor et al., 2006; Yurkov et al., 2015). Hawksworth, in 1996, proposed a practical definition for fungal species as 'groups of individuals separated by inheritable character discontinuities and for which assigning a species name is useful.' Because these discontinuities are best identified through combined genetic and phenotypic analysis, this definition aligns with the phylogenetic taxon species concept (Quaedvlieg et al., 2014; Wilkins, 2018; Vinarski, 2019).

Fungi have complex evolutionary paths, and their levels of genetic coherence, reproductive isolation, and phenotypic uniqueness often differ. Thus, a universal method for defining species remains elusive, complicating identification (Lücking, 2019).

Fungal diversity in reproductive modes and life cycles further complicates the definition of species boundaries. Morphologically distinct forms might reflect different life stages, commonly the sexual and asexual forms of the same species (Kendrick, 1979; Covert et al., 2007). In lichens, morphologically similar species pairs may either represent a single taxon or exhibit intricate evolutionary relationships where reproductive features are unreliable for differentiation (Mattsson & Lumbsch, 1989; Messuti et al., 2016). Moreover, a single lichen-forming fungus may show different vegetative forms depending on its algal partner, leading to distinct 'photosymbionts' (Armaleo & Clerc, 1991; Moncada et al., 2013).

The biological species concept, which focuses on reproductive isolation, is difficult to apply in fungi. Though early classifications of *Neurospora* included mating compatibility, fungal reproductive systems are often cryptic. Mating involves microscopic structures and rarely occurs in natural or laboratory settings. Failed mating experiments don't necessarily indicate species boundaries, as biotrophic fungi need a compatible host, and its absence might cause reproductive tests to fail (Cai et al., 2011). On the other hand, mating can sometimes occur between genetically and structurally distinct species through interbreeding. Many fungi appear to have lost sexual reproduction altogether, though some can be coaxed into it under lab conditions.

2.2 COX 1

Cytochrome c oxidase, or COX 1, is a unique barcode that is commonly used to identify different animal species using DNA. It's a protein that plays a crucial role in the energy production process within a cell. In humans, the instructions for making this protein are encoded in a gene called *MT-CO1*, which is found in mitochondrial DNA. Researchers sequenced a specific part of the COX1 gene, known as the barcode region, for 360 strains of mold from the *Penicillium* genus. They used PCR technique and special primers designed from fungal mt-genome in the *Trichocomaceae* family.

The COX1 gene provided unique identification codes (barcodes) for about 66% of the species studied. This was better than the ITS gene (approximately 25%) but not as effective as other protein-coding genes. Fortunately, the presence of introns (non-coding regions) in the COX1 gene was rare, occurring in only about 1% of the strains. When introns were present, they did not interfere with the DNA amplification process and had little effect on the usefulness of COX1 for barcoding purposes. Comparable results were also observed in a study examining seven species within the soil-dwelling genus *Leohumicola.*

Similar success in identifying species was achieved using both COX 1 and ITS genes. This was made possible by newly designed COX1 primers that targeted a specific subgroup, *Pezizomycotina*. Specifically, introns were detected in only 5% of the strains, indicating a relatively low frequency of occurrence. Previous attempts to utilize COX1 to identify mold species have been unsuccessful. For example, Geiser et al. (2007) found that COX1 did not provide enough variation to distinguish between different species within the *Aspergillus niger* complex, which is closely related to *Penicillium.*

Although not explicitly addressed in the text, the published phylogram revealed evidence hinting at the presence of multiple copies of the COX1 gene in certain species. Gilmore et al. (2009) found that individual strains of the *Fusarium* species had numerous copies of the COX1 gene, known as paralogs. It was evident from the presence of dual bases at certain positions or varying sequences among the copies. Unfortunately, these identical barcode sequences were shared among different species within the same genus, making it difficult to distinguish between them. The COX1 gene was found to be an effective barcode marker for distinguishing between species within the oomycete genera *Pythium* and *Phytophthora*, but with some unexpected issues (Martin & Tooley, 2003). Low copy numbers characterize COX1 genes.

DNA barcoding is a rapid, precise, and efficient technique for species identification, offering a reliable means of distinguishing between different species. DNA barcoding has emerged as a potent means of species discrimination in recent years, enabling accurate and efficient identification of species. Scientists can leverage the transformative power of uncapped DNA barcoding to overcome existing and emerging challenges, capitalizing on its innovative approach to drive progress in the field.

Becker & Pusharava (2023) underscore the efficacy of DNA-based diagnosis for species identification. Their findings emphasize the importance of the mitochondrial cytochrome c oxidase gene as a key component in global identification system, facilitating a deeper understanding of biodiversity. The study revealed that COX1 sequence divergence is a reliable marker for taxonomic classification, enabling the accurate assignment of organisms to their corresponding taxonomic ranks, and providing a higher resolution than morphological characterization alone. Researchers conducted a comprehensive review of DNA barcoding, highlighting its vast potential in various disciplines, including eco-informatics, community ecology, invasive species management, macro-evolutionary studies, trade and evolution analysis, food web dynamics, species interaction analysis, and spatial ecology. This underscores the significant contribution across multiple fields of study.

Fungi pose a significant limitation for multidisciplinary research in ecological and biodiversity studies. DNA barcoding utilizes a standardized DNA sequence of

500–700 bp to identify and distinguish species across all kingdoms of life using broadly applicable primers that span the diverse taxonomic groups. Reference barcodes should be sourced from adeptly articulated voucher specimens forwarded in accredited natural assemblage with accompanying validation through publicly available sequence chromatograms.

For effective DNA barcoding, interspecific variation should surpass intraspecific variation, with a clear barcode gap separating species. Ideally, a barcode sequence should be constant within a species and unique among different species. A universal barcode locus, applicable across all kingdoms of life, is optimal. As a standard barcode for animals, the mitochondrial gene encoding cytochrome c oxidase subunit 1 serves this purpose and has been adopted by the Consortium for the Barcode of Life (CBOL) as the default marker for all organisms, including fungi, in the Barcode of Life initiative. In the fungal kingdom, the ITS region is traditionally used for identification by mycologists, but the default COI marker is more reliable in distinguishing closely related species within certain clades. However, in plants, CO1 has limited utility for species differentiation. Consequently, a two-marker system was adopted, utilizing portions of the *matK* gene and *RBCL* gene. The CO1 gene functions adequately as a barcode in certain fungal genera such as *Penicillium*. Here, reliable primers provide sufficient species resolution (67%) in younger lineage.

However, the outcomes from different examined groups are self-contradictory, often necessitating further biological research. Furthermore, degenerate primers designed for broad applicability across *Ascomycota* are challenging to evaluate, as expansion occurrence might not necessarily indicate priming conflicting. The distinct discrepancy arises from the inconsistent presence of multiple introns. Fungal genomes often contain multiple copies of genes with varying lengths and sequences. Additionally, identical gene sequences are sometimes shared among different species, notably certain fungal lineages, such as the early-diverging group *Neocallimastigomycota*, which comprises obligate anaerobic, zoosporic gut fungi that lack mitochondria altogether. Ultimately, due to the microscopic and inconspicuous nature of most fungi, as well as the challenge of culturing many species, the development of a robust, universal primer is crucial for detecting a truly representative profile of fungal diversity. Nuclear ribosomal DNA, used for mildew recognition and phylogenetics for over a score, is now considered a promising substitute for CO1.

2.2.1 Rise of ITS Region as Preferred DNA Barcode Marker

The COX1 gene has limited utility as a barcode due to its restricted interspecific variability in certain groups. In contrast, the ITS region, situated between structural RNA sequences on a shared precursor transcript, exhibits greater variability, making it a more suitable barcode marker. The ITS region is a non-coding RNA segment, positioned 5' to 3' between functional RNA sequences.

The ITS region is a publicly employed genetic marker in classification and biological evolution due to its ease of amplification, even from minute DNA quantities, courtesy of its high copy number. Furthermore, the ITS region exhibits substantial variation, allowing for differentiation even between closely related species. The majority of mycologists anticipate that the ITS region of nuclear ribosomal DNA will be established as the primary barcode for the kingdom of fungi because its popularity stems from its utility.

2.2.2 Exploring Alternative Novel Barcode Regions for Fungal DNA Barcoding Analysis

Besides the ITS region, several alternative barcode regions have been explored for fungal identification, including:

- Cytochrome oxidase 1 (COX1)
- Large subunit (LSU) ribosomal DNA
- Histone H3 spacer region
- Enolgation factor 1-alpha (EF-1) gene
- RNA polymerase subunit genes (RPB-1, RPB-2)

As sequencing costs continue to decline, the feasibility of sequencing longer DNA regions increases. In the future, sequencing larger genomic regions may become desirable, as it could provide:

- More comprehensive phylogenetic information
- Increased resolution for species identification
- Enhanced understanding of evolutionary relationships

2.3 INTERNAL TRANSCRIBED SPACERS

The eukaryotic rRNA sequence comprises basic genes (18S, 5.8S, 28S); these genes are co-duplicated by RNA polymerase I as a single regulatory unit, which is then processed into individual rRNA molecules. Following transcription, the polycistronic RNA undergoes processing, where two internal transcribed spacers (ITS) are removed. These spacers are excised, resulting in mature rRNA molecules. The ITS region is often marked for phylogenetic analysis and species recognition.

SSU: The 18S nuclear ribosomal small subunit (SSU) gene is a widely used marker in phylogenetics. Its homolog, 16S, serves a similar purpose but is more commonly used for species-level diagnostics in bacteria. Both genes provide valuable information for understanding evolutionary relationships and identifying organisms.

LSU: The 28S nuclear ribosomal large subunit (LSU) gene is often utilized for species discrimination by providing distinct species-level resolution, or in combination with other genetic markers, such as ITS or SSU, to enhance phylogenetic accuracy and species identification. For yeast, the 28S nuclear ribosomal large subunit (LSU) gene has been used for species characterization since the advent of DNA barcoding. This region has a proven track record for distinguishing yeast species, making it a valuable tool for identification and classification. Contributions from 500 research journals indicate that data derived from 11,500 scientific studies, encompassing 2,500 genera and 15,500 species, are linked to 56% of Latin binomial names. The GenBank database currently contains around 772,000 full-length fungal ITS sequences. A significant portion of unclassified sequences are from environmental samples.

In some environmental studies, researchers have used a combination of ITS and LSU markers. Additionally, ITS is used in certain fungal groups to estimate species boundaries by measuring genetic distance, providing a preliminary indication of species delimitation. Cladistic perspectives are increasingly being applied to spot grouping units in surrounding fungal samples, family analysis, and species identification, particularly for *Ascomycota* and *Aspergillus*. These genes generally outperform rRNA genes in resolving relationships at various taxonomic levels, providing a more accurate understanding of fungal evolution and diversity. Specialized identification databases often employ multiple markers for accurate species identification. Examples include translational elongation factor 1-alpha (TEF-1 alpha) for *Fusarium* species, and beta-tubulin for *Penicillium* species.

However, a notable limitation is the lack of standardization across these databases, which can hinder comparison and consistency in identification results. Due to its ubiquitous presence in fungi, single-copy nature, and relatively slow rate of sequence divergence among protein-coding genes, the largest subunit of RNA polymerase II (RPB1) shows promise as a potential fungal barcode. The mentioned characteristics make RPB1 an attractive candidate for distinguishing fungal species and reconstructing phylogenetic relationships. The phylogenetic utility of RPB1 has been demonstrated in studies of *Ascomycota*, *Zygomycota*, and *Mycosporidia*. RPB1 primers were also developed for the Assembling the Fungal Tree of Life (AFTOL) project, which included these genes in its subsequent analysis. However, its potential as a barcode marker remains untested and unexplored. This chapter presents a collaborative, international effort and multi-laboratory attempts to create a standardized DNA barcode for the kingdom of fungi, focusing exclusively on fungal organisms. The evaluated comparative performance of barcoding regions (ITS, LSU, SSU as nuclear regions) is discussed.

2.3.1 PCR Success

PCR amplification of rRNA genes proved to be more reliable across various fungal species compared to protein-coding markers. The success of PCR amplification varied among different taxonomic groups; for instance, the ITS PCR amplification success rate ranges from 100% to 65%, indicating inconsistent performance across different fungal lineages. The success rates of other markers were comparable. However, RPB1 amplification success varied significantly, ranging from 80% to 14%.

In contrast, ITS amplification was generally more straightforward, with approximately 80% of respondents reporting no issues with PCR amplification. Ribosomal markers had favorable issues with PCR expansion compared to protein-encoding guide monitors. Based on overall performance and species discrimination, the SSU gene has significant limitations as a barcode marker due to the following reasons. This is because SSU's inability to distinguish between species is due to high sequence similarity. This is also because SSU's words combined performance criteria index (PCI) makes it less effective for species finding. In contrast, other markers, such as LSU and ITS, show more promise for fungal identification.

LSU exhibits a large barcode gap, making it more suitable for species discrimination. ITS is commonly used for fungal identification due to its variability and effectiveness for species discrimination.

As a result, SSU can be eliminated as a candidate locus for fungal barcoding, while LSU and ITS are considered potent barcodes for accurate species identification. The LSU gene, a popular phylogenetic marker among mycologists, demonstrated exceptional performance with virtually no issues with PCR amplification, alignment, sequencing, and editing.

A significantly superior barcode gap compared to the SSU gene suggests that ITS is the most effective marker for fungal species discrimination. The overall likelihood of right species recognition through ITS is analogous to the completion rate. Mushrooms, with a slightly lower success rate, may be encountered as essential mold microbe groups within the Ascomycota (e.g., molds), which are fungi used in food and agriculture.

ITS was a strong contender to RPB1 and, to a great extent, sampled protein encrypting guide monitor in terms of species identification success. However, ITS has a critical advantage: its significantly higher PCR exaggeration work completion rate makes it a more practical and reliable barcode marker. The ITS primer used depicts the broad applicability across various fungal lineages, with several primers functioning as nearly universal primers, able to amplify ITS regions from a wide range of fungi.

However, overall primer sets exhibit some degree of bias, and a practical solution is to employ multiple primer combinations to ensure more comprehensive and accurate results. Considering the cumulative evidence, we proposed the ITS region as the standard DNA barcode for fungi, given its advanced species discrimination power, broad applicability across fungal lineage, and high PCR amplification success rate. This proves ITS is the universal standard for fungal identification, facilitating more accurate and efficient research in mycology.

While the proposal to establish ITS as the standard fungal barcode will likely satisfy most fungal biologists, it may not meet the needs of all researchers due to the fungal kingdom's ancient origin and vast hereditary variability; it is unrealistic to expect a unified marker barcode system to accurately identify every specimen or culture to the species level.

2.3.2 ITS Is a Powerful Tool

Despite overall challenges, ITS remains a powerful tool for fungal DNA barcoding, offering full resolution to distinguish more inter-related family members of this species and a high PCR and sequencing success rate. Beyond fungi, the ITS region may also serve as a viable barcode for other organisms. The prospect of multi-kingdom analysis of complex ecosystems, such as soil and high-copy-number ITS markers, aligns with the novel region of DNA barcoding. This approach is already showing promise, enabling the simultaneous amplification of fungi and other eukaryotic species from soil samples.

Protein encrypting genes are widely used as cladistic identifiers, and they have been employed as DNA barcodes with limited taxonomic scope in various fungal groups, serving as effective identifiers for the specific success of fungi. While protein markers offer higher species resolution, their susceptibility to PCR non-accomplishment renders them impractical as a universal barcode for the vast and diverse kingdom of fungi, which encompasses a broad phylogenetic scope. Certain kingdom-wide PCR expansions need to be evaluated for other commonly used protein encrypting identifiers, including (1-alpha, beta-tubulin, and actin), to determine their suitability as the universal barcodes for the fungal kingdoms. In conclusion, the advantage of utilizing the ITS region as a biological identifier and fungal DNA barcode is that the total ribosomal gene cluster is organized in the form of microsatellites, resulting in aggregate replicas, which enhances its utility and effectiveness. Specialized tools have been developed to identify specific fungi, such as *Trichoderma* species, using analysis of ITS and TEF 1-alpha sequences with programs like Trichokey or TrichoBlast. The effectiveness of this approach in identifying human pathogenic fungi has been demonstrated in studies, including those by Erickson et al. (2017), which highlights the utility of these tools in fungal identification.

2.3.3 Limitations of ITS

Furthermore, the limitations of ITS sequencing in identifying species within certain fungi groups, as well as the failures of universal ITS primers in a subset of other groups, will require thorough documentation to ensure a comprehensive understanding of the marker's strengths and weaknesses. Shared ITS sequences among different species have already been recorded in species-rich genera characterized by truncated amplicons, including money-saving essential fungi such as *Cladosporium*, *Penicillium*, and *Fusarium*. Similarly, in certain genera, ITS sequences are indistinguishable among respective complexes of critically important species.

While the ITS region shows promise as an effective DNA barcode in various lichen-based lineages, the allelic float may hinder lineage grouping of heritable polymorphisms in certain slowly emerging classes, potentially limiting its utility for species discrimination. The selection of primers for ITS amplification can introduce bias for particular taxonomic groups. For instance, the 'universal' ITS primers have been shown to fail to amplify approximately ≈10% of the tested fungal specimens, highlighting the need for careful primer choice to ensure representative coverage (Yurkov et al., 2015).

During sequencing, the intragenomic ITS heterogeneity can result in overlapping reads of varying lengths, potentially generating unreadable chromatograms due to superimposed single traces. Moreover, the anti-sense nature of the ITS domain can introduce real indel variation, rendering it challenging to systematically line up ITS sequences from extremely radiating taxonomic categories, thereby limiting the utility of ITS for large-scale phylogenetic analysis.

The tandemly recurrent nature of the ribosomal cistron flock can lead to substantial intragenomic ITS sequence heterogeneity, resulting in variability among paralogous copies within individual fungal genomes and thereby complicating molecular phylogenetic and barcoding analyses. The standard method of generating ITS nucleotide-block DNA base-pair metabarcoding of climate samples, which often contain a diverse array of fungal species, involves a multi-step process: initial

PCR amplification, cloning of the resulting amplicons, and subsequent sequencing of the cloned PCR products. Conversely, morphologically well-defined yet evolutionary interconnected species may exhibit minimal or subtle differences in their ITS sequences, often limited to a few nucleotide variations. Sole reliance on ITS barcode data for recognition of such species can hide the genuine variety and potentially result in misidentification, underscoring the need for a complementary analysis of morphological and ecological characteristics, or comparison with additional diagnostic genetic markers, to ensure accurate taxonomic classification.

Variations within the genome, such as the presence of dual paralogs or nonparalogs within the spore-producing structures, seeds, or both *Ascomycota* and *Basidiomycota*, may result in the overestimation of intraspecific variation.

It is evident when ITS distances are comparable between the following comparisons:

- Interspecies comparison
- Intraspecies comparison
- Intergeneric comparison

This similarity in distances highlights the challenges of using ITS alone for species delimitation.

In an extensive data set of ITS sequences from fungal DNA barcoding databases, the highest levels of variation were predominately observed in early-diverging fungal lineages. This finding underscores the reality that fungal diversity, particularly in early-diverging lineages, remains under-sampled and underrepresented in DNA sequence databases. Therefore, it is most likely that flooding fluctuation reflects the existence of many cryptic species, highlighting a crucial area for further research and investigation.

2.3.4 Low-Resolution as a Secondary Barcode

An increasing body of research has questioned the effectiveness of the ITS domain for species borderline, recognition, and identification across certain fungal lineages. One limitation is its accessional failure to amplify in some taxa; however, overall sequencing success for ITS remains higher compared to many alternative genetic markers (Schoch et al., 2012).

More critical limitations of the ITS region include its insufficient discriminatory power and the potential occurrence of non-homologous ITS copies within a single genome. ITS has been shown to lack adequate resolution for distinguishing closely related species in several groups, including indoor and food-associated molds, such as *Aspergillus* and *Penicillium*, as well as plant and animal/human pathogens like *Alternaria*, *Cladosporium*, *Collectotrichum*, *Fusarium*, and members of the Oomycota, such as *Phytophthora*. Similar challenges have been observed in other fungal groups, including freshwater Sordariomycetes, *Trichoderma*, and slime molds.

For these cases, alternative genetic markers—commonly referred to as secondary barcodes—have been proposed to improve resolution. These include the intergenic spacer (IGS), B-tubulin II (TUB2), the largest (RPB1) and second-largest (RPB2) subunits of DNA-directed RNA polymerase II, translational elongation factor 1-alpha (TEF1), DNA topoisomerase I (TOP1), phosphoglycerate kinase (PGK), and the mitochondrial cytochrome c oxidase subunits I (COX1) and (COX2). Interestingly, even in taxonomic groups where the ITS region generally performs well, exceptions have been noted. For instance, the macrolichens *Sticta fuliginosa* and *S. limbata*, both widespread and of conservation concern, were found to be indistinguishable using ITS, suggesting that this issue may not be limited to specific taxa. Instead, it likely reflects rapid or ongoing evolutionary processes, which, although more frequently observed in certain fungal groups, can potentially affect any lineage. Recent barcode analysis further indicates that approximately 6–17% of yeast and filamentous fungal species cannot be reliably distinguished using ITS alone.

Given the limitation of ITS as a functional barcoding DNA in fungus, the need for a secondary barcode locator became apparent. Consequently, many efforts have been made to launch alternative hereditary identifiers that can serve as supplementary DNA barcodes, mirroring the approach used in plant DNA barcoding, where *matK*, RBCL, and nuclear-ITS are often applied in combination to achieve more robust identification. In cases where ITS data lack sufficient resolution to determine the phylogenetic position of an unknown or poorly resolved species, it is recommended to first determine the isolate's identity using LSU, SSU, or a combination of both, which can provide a broader phylogenetic context and help place the isolate within a larger taxonomic framework (Goulding & Dayrat, 2016).

The ITS region, although widely used, has been subject to scrutiny due to its limited effectiveness in differentiating interconnected species of fungi (Nilsson et al., 2008; Kruse et al., 2018a). ITS occasionally fails to amplify in some lineages, although its overall amplification success remains higher than that of many alternatives (Schoch et al., 2012).

More importantly, ITS often lacks the declaration needed to distinguish interconnected species of fungi. This is evident in genera like *Aspergillus*, *Penicillium*, *Alternaria*, *Cladosporium*, and *Fusarium*, as well as Oomycetes such as *Phytophthora*. Alternative markers, such as IGS, TUB2, RPB1, RPB2, TEF1, and mitochondrial genes like COX1 and COX2, are frequently used (Geiser et al., 2007; Meyer et al., 2019).

Notably, even within groups where it typically works well, exceptions exist. For instance, *Sticta fuliginosa* and *S. limbata*, despite being morphologically and ecologically distinct, are not distinguishable using ITS alone. This may be due to rapid evolutionary changes that ITS variation has not yet been able to capture. Barcode studies estimate that 6–17% of fungal species cannot be confidently resolved using ITS data alone.

Meyer et al. (2019) found that up to 25% of fungi pathogenic to humans or animals could not be reliably identified with ITS. Similar gaps exist in plant-pathogenic groups within Dothideomycetes and Sordariomycetes (Damm et al., 2012). On the other hand, certain lichen-forming fungi within Dothideomycetes, such as *Strigula*, exhibit high resolution with ITS (Jiang et al., 2016–2020).

These discrepancies suggest that the reliability of ITS is linked to life history traits and ecological strategies, such as host specificity, which may drive rapid speciation not reflected in ITS divergence. Thus, more nuanced approaches are needed, particularly for taxa of economic or medical importance, which are more frequently sampled and studied in detail.

2.3.5 Future Perspectives on Applications of ITS

2.3.5.1 Role Performed by ITS Region in Fungal Community Studies

Currently, the ITS domain is considered a recognized biological marker for large-scale fungal community profiling. Although researchers have evaluated alternative gene regions such as *tef1*, *rpb1*, and *rpb2* as substitutes, these often pose challenges due to their high sequence variability, particularly in the third codon positions, which makes it difficult to design broadly applicable primers across divergent groups of fungus species (Stielow et al., 2015). Despite its broad use, the ITS region has limitations, including variable resolution at the species level, intragenomic differences, and the absence of a distinct barcoding gap. Furthermore, technical artifacts introduced during PCR, such as chimeras or amplification bias, can complicate the interpretation of high-throughput sequencing datasets. Interestingly, its multicopy nature, which can introduce noise, also enhances its detection sensitivity, highlighting the balance between universality and specificity.

The extent to which technical biases and molecular limitations affect outcomes often depends on the ecological context. For example, beta diversity—community variation across environmental gradients—tends to remain stable despite differences and sequencing errors in data processing. Botnen and colleagues demonstrated that fungal community structure is primarily determined by the dominant operational taxonomic units (OTUs) that remain consistent regardless of the pipeline choices.

Even when rare OTUs were excluded and only abundant taxa retained, overall community patterns persisted, suggesting that some shortcomings of the ITS region may have limited effects in practical ecological applications. Moreover, when studies are conducted at local scales, the issue of intraspecific ITS variation tends to diminish.

In many ecological investigations, identification at the genus level is sufficient, making precise species-level resolution less critical. However, studies focused on richness or diversity metrics are more sensitive to the chosen analytical methods, as the resulting OTU counts may vary widely. PCR primer mismatches and variable amplicon lengths can also skew detection, resulting in biased representations of certain lineages.

2.3.6 Fungal DNA Barcoding: Applications and Utilization

2.3.6.1 Ensuring Food Protection

A study by Raja et al. investigated the accuracy of labeling on mushrooms intended for consumption using ITS region DNA barcoding. The analysis of 33 samples, including dried mushrooms, mycelium powder, and capsules comprised of dietary supplements, revealed significant discrepancies. Only 30% of the samples had correct binomial fungal names on the product labels.

2.4 FUNGAL PATHOGENS

Studies by Vi Hoang et al. (2019) demonstrated that ITS sequences can be used to reliably identify commercially valuable species and their counterfeits at the species level. Researchers investigated the quality of identifying disease-causing fungi using primary ITS and secondary TEF-1 alpha barcode markers. The results showed that for certain genera, including *Candida diutina* and *Pichia* species, identification was straightforward using either ITS or TEF-1 alpha alone, as well as with the combination of both markers. During the study of common pathogenic *Candida* species, ITS failed to distinguish some of their members. In contrast, the TEF-1 alpha marker accurately identified all investigated species within the clade.

2.4.1 How DNA Barcoding Was Performed: Step-by-Step Guide

Step 1:

Creating reference library:

The process begins with creating a reference sequence for different species, which is then stored in a global library with public access. This library enables researchers to compare and identify unknown barcode sequences.

The importance of accurate sample identification:

Proper identification and maintenance of the reference sample are crucial to ensure the accurate labeling of the reference barcode sequence. Incorrect identification can lead to incorrect labeling, making it impossible to identify unknown species.

Step 2:

- DNA extraction, amplification, and sequencing.
- DNA extraction: Using commercial kits for high-speed extraction.

Barcode region amplification involves replicating barcode sequences using a scientific method and sequencing the replicated sequences.

Step 3:

Data management and sharing:

The final step involves depositing the DNA barcode sequences and reference sample information into a public database, facilitating data access and comparison.

Global collaboration:

The CBOL collaborates with organizations worldwide to implement joint DNA-barcoding projects, currently working with 50 countries to advance the field of biodiversity research.

2.4.2 Molecular Taxonomy: A New Era in Fungal Identification

The use of morphological features in classifying, identifying, and understanding the taxonomic relations of living organisms has a rich history spanning over many hundreds of years. This traditionalist approach, known as morphological identification, has long been the cornerstone of species identification in scientific research. The morphological strategy is based on the idea that distinct morphological characteristics can distinguish individuals at the species level.

This concept of species has two key components: (1) individuals within a species exhibit consistent and homogeneous morphological traits and (2) these traits show clear and distinct differences. The morphological concept of species has two significant strengths: its broad applicability to any fungal taxon, and its extensive and well-established history of use (Termorshuizen and Arnolds, 1997a). By custom, the identification of fungi has relied on phenotypic features, including morphological and physiological properties. However, these characteristics, particularly the morphology, are significantly leading to high variability (Raja et al., 2017).

The description and identification of fungal species typically involve the consideration of multiple morphological features, as differences in these characteristics can be used to distinguish between species of fungal kingdoms (Taylor et al., 2000; Crous et al., 2009). Another notable limitation of relying on morphological characteristics for fungal identification is that the number of recognizable traits is often far fewer than the number of known species within a genus, or the number of distinct species that exist.

Mycologists have encountered numerous challenges when relying on morphological data for identification, particularly with *Fusarium* species. A significant issue is that this fungus exhibits limited morphological features, making it difficult to distinguish between species (Termorshuizen and Arnolds, 1997b). The advent and application of molecular techniques have sparked controversy in the taxonomy of *Fusarium* species. Specifically, the combination of phylogenetic species concepts with DNA collection has resulted in the identification of new organism classes, including several sibling species (Hebert et al., 2003).

For instance, within the *Fusarium graminearum* complex, 15 distinct phylogenetic species have been identified, yet these species are morphologically indistinguishable, highlighting the limitations of relying solely on morphological data for species identification (Arve et al., 2011). That limitation is also evident in black yeast fungus, which refers to melanin-producing fungi that exhibit both yeast and mycelial forms in culture, and are known to be pathogens of humans and plants, particularly affecting the brain and other organs in immune-compressed individuals, with potentially fatal consequences.

While black yeast is now contributing to the deterioration of ornamental stones, it also plays a beneficial role in environmental bioremediation, particularly in the biodegradation of oil pollutants, highlighting their dual impact on the environment (Badali et al., 2011).

The challenges associated with structure-based findings directed scientists to adopt a biological approach, notably the polymerase chain reaction, which enables the replication of a species' gene region from DNA samples (Crous et al., 2009;

Nancy & Binnicker, 2009). By extracting DNA from fungal cultures, researchers can now amplify and analyze parts of the genome for diagnostic purposes, overcoming the limitations of traditional identification methods. The widespread adoption of the PCR technique has enabled the direct discovery and measurement of fungi from various samples, including infected plants, blood, air, soil, and other environmental sources, revolutionizing the field of fungal diagnostics (Nancy & Binnicker, 2009). Molecular approaches offer significant advantages over traditional methods, including faster processing times, higher accuracy, and increased sensitivity. Additionally, these methods reduce the health risks associated with handling contaminated samples, providing a safer working environment for researchers and laboratory personnel (Nancy & Binnicker, 2009).

Currently, nucleotide-sequenced data play a crucial role in fungal identification across various fields, including medicine, agriculture, and commercial industry. Furthermore, this information is also used to investigate the kinship between living things at specific classification levels, contributing significantly to our understanding of evolutionary biology.

Fungi are among the most extensively studied groups, with a substantial number of fungal pathogens affecting humans, plants, and industrial applications having had their genomes fully sequenced. The exploration of fungal DNA barcoding began in 2003, with researchers at the University of Guelph in Canada taking the lead. The following year, the CBOL made notable advancements in establishing a more reliable identification tool through the creation of a comprehensive DNA barcode reference library (Nilsson et al., 2008).

2.4.3 Inferring Reproductive Isolation and Modern Approaches in Fungal Taxonomy

Since directly measuring reproductive isolation in fungi is often impractical, researchers commonly use indirect methods such as the Genealogical Concordance Phylogenetic Species Recognition (GCPSR) framework. This method identifies consistent genealogical patterns across multiple genetic loci as indicators of evolutionary independence between lineages (Taylor et al., 2000). Although GCPSR has proven effective in fungi, it doesn't necessarily reveal the presence of intrinsic reproductive barriers and may instead reflect patterns of population structure rather than distinct species boundaries.

Another approach involves examining mating-type genes to infer reproductive compatibility among fungal taxa. In many cases, reproductively incompatible and genetically distinct lineages also align with phenotypic classifications, as seen in genera such as *Neurospora*, *Cryptococcus*, and *Fusarium*. However, focusing too heavily on Mendelian traits can sometimes lead to discrepancies between phenotype-based and DNA-based species definitions (Aime et al., 2018).

To address these challenges, modern fungal taxonomy increasingly relies on genealogical frameworks. These methods include phylogenies derived from single or concatenated gene sequences, genealogical concordance analysis, and phylogenomic approaches. One of the strengths of this framework is that it allows species hypotheses to be tested explicitly, with phenotypic traits later interpreted in light of evolutionary history.

Additionally, molecular approaches provide access to a much larger dataset, while morphological matrices may include a few hundred or fewer observable traits. In contrast, DNA-based datasets can range from hundreds of nucleotide positions in single-gene analysis to thousands or even hundreds of thousands in multilocus or genome-wide studies. Despite this, reconstructing phylogenetic relationships still poses difficulties due to the complex and sometimes poorly understood nature of evolutionary processes.

As a result of these complexities, contemporary fungal taxonomy integrates data from single-locus and multilocus phylogenies, genealogical concordance, and genome-scale analysis. This strategy benefits from a rigorous evolutionary framework, allowing retrospective assessment of phenotypic traits. The richness of molecular data significantly surpasses that of traditional phenotypic matrices: whereas morphological datasets are limited in scope, genomic data span from hundreds of nucleotide positions in single-gene studies to hundreds of thousands in phylogenomic datasets. Nevertheless, unresolved evolutionary questions can still limit the clarity of phylogenetic reconstructions.

To improve taxonomic resolution and overcome limitations in statistical power and gene sampling, whole-genome sequencing is increasingly applied in fungal systematics. In bacterial systematics, methods like Average Nucleotide Identity (ANI) have become popular due to their efficiency, alongside maximum-likelihood phylogenetic tools. A large number of prokaryotic genomes, including those from type strains, are now publicly available. Furthermore, genome-based metrics such as the Percentage of Conserved Proteins (POCP) are increasingly used to differentiate closely related prokaryotic species.

The development of effective identification tools remains a core responsibility for taxonomists across all biological disciplines, including mycology. Utilizing current phylogenetic frameworks, specialists are encouraged to integrate cutting-edge methodologies to compile comprehensive datasets of diagnostic characters. To ensure accessibility and reproducibility, both identification tools and taxonomic descriptions of novel species should be made freely available. Registration of fungal taxa in repositories such as MycoBank, Index Fungorum, or Fungal Names secures formal nomenclature-based recognition, with the deposition of illustrative material, although not mandatory, strongly encouraged. Challenges associated with open-access publishing, including publication fees and impact considerations, may be alleviated by archiving reprints in open repositories like bioRxiv, enabling unrestricted access while preserving citation of the peer-reviewed source.

The implementation of unified digital protologues with semantic standardization offers the potential for automated aggregation, organization, and analysis of taxonomic data at both the specimen and species levels. However, this approach is hindered by terminological inconsistencies and the extensive character sets necessary to represent the full breadth of fungal diversity, of which only subsets are generally relevant for specific lineages.

2.4.4 Verification

Despite the availability of identification tools, taxonomic assignments are often accepted without critical scrutiny, which can lead to misidentifications due to limitations in algorithm performance or inaccuracies in reference databases. These errors can occur in both morphological and molecular approaches, particularly when sequences are mislabeled, aligned with outdated or inconsistent taxonomies, or when pairwise similarity methods such as BLAST are used without rigorous interpretation (Nilsson et al., 2008). The use of different BLAST algorithms—megablast, discontinuous megablast, or blast—can produce varying results depending on sequence length, alignment parameters, and whether full or partial barcode regions (e.g., ITS1 or ITS2) are submitted. These factors underscore the importance of meticulous verification procedures, which help prevent circular reasoning and ensure accurate taxonomic assignments.

In high-throughput contexts such as environmental metabarcoding, full verification of all identifications is often impractical, necessitating trade-offs between processing efficiency and taxonomic precision. The type of diagnostic characters, whether phenotypic or molecular, determines the appropriate verification pathway. Phenotypic identifications typically require consulting original descriptions and type material, which may be accessed physically or via digital repositories like JSTOR Global Plants or the Mycology Collections Portal.

Phenotype-based verification, while sometimes overlooked, remains a vital complement to DNA-based identification. Linking images of voucher specimens with associated sequence data deposited in accredited repositories can greatly enhance the reliability of identifications. Standardizing metadata for voucher annotations is also crucial, particularly during the registration, publication, and submission of sequence data to platforms like GenBank and the International Nucleotide Sequence Database Collaboration (INSDC). Innovative formats such as the AJOM Fungal Notes series promote the documentation of well-identified collections, especially those of previously described species with new molecular data. Such datasets, including images, are increasingly shared through online platforms dedicated to particular fungal groups.

Sequence-based validation of fungal identifications can be carried out through multiple sequence alignments and the construction of phylogenetic trees, which help determine the evolutionary placement of reference sequences, ideally those originating from type specimens. Although BLAST includes a feature that allows users to limit search results to sequences designated as originating from type material, the scarcity and uneven taxonomic representation of such sequences limit the broader applicability of this function in comprehensive fungal analysis. In situations where type-derived sequences are not available, researchers can utilize vetted repositories that house high-quality reference sequences, even if they are not directly linked to type material.

These resources offer rigorously curated sequence data that can significantly improve the accuracy and reliability of molecular taxonomic identifications, particularly for clinically or agriculturally important strain-level differences. In some instances, discrepancies in resolution between the ITS region and more variable genetic markers are addressed by recognizing intraspecific taxa. Conversely, in other cases, such as the multiple IGS-defined lineages within the widespread basidiomycete *Schizophyllum commune*, no formal taxonomic framework has been established. Consequently, identical phylogenetic patterns can lead to divergent taxonomic interpretations, typically influenced by practical requirements. Thus, the degree of resolution targeted by DNA barcoding should be context-dependent, irrespective of how much resolution is reflected in formal taxonomy.

In many instances, identification can be narrowed to a species complex within a specific clade, but resolution at the species level remains unattainable. In such cases, dual-marker barcoding systems, such as nuLSU/ITS or TEF1, have proven effective, particularly for yeasts and human or animal pathogens (Vi Hoang et al., 2019). However, implementing these dual-marker approaches in metabarcoding workflows continues to present significant challenges.

2.4.5 Challenges and Solutions in Pairwise Similarity Analysis

2.4.5.1 OTU Clustering

A primary limitation of DNA barcoding lies in the widespread reliance on pairwise similarity methods, typically conducted via BLAST searches or clustering algorithms such as USEARCH, VSEARCH, or MultiLevel Clustering. These techniques have gained popularity due to their ease of integration into automated analytical pipelines, enabling the rapid processing of extensive datasets with minimal manual intervention. However, unlike multiple alignment-based phylogenetic methods, pairwise similarity assessments can misinterpret positional variability, potentially leading to inaccurate delineation of taxonomic units and phylogenetic relationships.

For example, in a multiple sequence alignment, a variable indel region such as [AG], [A], or [G] will align [A] with either another [A] or a gap but not with [G]. In contrast, pairwise alignment would interpret [A] and [G] as a substitution. Although this distinction may seem negligible, it can significantly impact OTU clustering, particularly when such variability results from sequencing errors. Consequently, OTUs generated through clustering algorithms often differ in both quantity and composition from phylogenetically defined entities. This approach offers an alternative that minimizes stochastic effects on cluster formation and inflation.

Enhanced accuracy without compromising computational efficiency can be achieved through high-confidence OTU (hc-OTU) clustering, which utilizes homopolymer compaction techniques. The integration of tools such as PaPaRa during read processing significantly reduces sequencing errors before OTU clustering. Lücking et al. demonstrated that automatic correction of homopolymer-associated errors using PaPaRa led to a 94% improvement in OTU clustering precision. To further refine clustering outputs, the LULU package can be applied post hoc to eliminate potentially spurious OTUs. Nevertheless, clustering methods rely on predefined similarity thresholds for grouping sequences, yet such fixed thresholds lack a universal basis for accurately delineating fungal species.

Optimal similarity thresholds vary across fungal lineages, as demonstrated by analysis of two extensive barcode datasets. When the selected genetic marker lacks sufficient resolution, even applying the highest similarity thresholds fails to produce reliable OTU estimates. Commonly, clustering pipelines default to a 97% similarity threshold, while curated reference databases, such as UNITE and the ISHAM ITS barcode collection, employ a more stringent 98.5% threshold for defining 'species hypotheses' based on ITS sequences. The use of fixed pairwise similarity thresholds can lead to an underestimation of species richness, while OTU clustering bias may simultaneously result in the overestimation of taxonomic units. Consequently, some OTUs may not correspond to actual taxonomic entities, whereas genuine taxa might be overlooked. This issue becomes more complex in lineages exhibiting substantial ITS sequence heterogeneity, occasionally exceeding 10%. Additionally, arbitrary adjustments to predefined similarity thresholds, ranging from 97% to 98.5%, can significantly influence the detection of taxonomic entities in clustering-based analysis.

2.4.6 Advantages of DNA Barcoding: Accessibility, Efficiency, and High Throughput

DNA barcoding presents a rapid, cost-effective, and scalable method for identifying biological diversity compared to traditional morphology-based taxonomy. When applied to large-scale studies, barcoding can reduce per-sample costs to as low as $0.10, provided workflows are efficiently scaled. However, the cost-effectiveness and utility of barcoding depend on factors such as the taxonomic group of interest, study goals, and regional access to sequencing infrastructure. In areas with limited technological resources, morphological identification may remain a more feasible option. To address this, partnerships have emerged that combine local taxonomic expertise with centralized sequencing hubs. Expanding barcoding initiatives globally must prioritize equitable access to technologies and fair benefit-sharing, aligning with international agreements such as the Nagoya Protocol.

The advent of high-throughput sequencing (HTS), whole-genome sequencing (WGS), and metagenomics has significantly reduced the cost of sequencing. Although these methods can provide more detailed genomic data, they also require greater computational resources, data storage, and bioinformatics expertise than traditional DNA barcoding. The strengths of barcoding and genomics are complementary—barcoding enables species-level identification across many samples, while genomics provides insights into deeper evolutionary and functional processes.

2.4.7 Curated Databases: Cornerstones of Reliable Barcoding

Databases such as GenBank and BOLD (Barcode of Life Data Systems) play a central role in storing and sharing DNA barcodes. However, their accuracy can be compromised by issues such as misidentification, sequencing errors, or contamination (e.g., co-amplification of non-target DNA). Although quality control measures are in place, such as compliance checks and metadata integration in BOLD, errors can still occur at any stage of the workflow.

For barcode-based research to remain effective, reference libraries must be regularly updated and curated. Incentivizing researchers to contribute to and maintain public datasets through funding or academic recognition could enhance the quality of databases. Barcoding specimens from natural history collections and including barcodes in new species descriptions, when feasible, can further enhance reference coverage and taxonomic resolution.

2.4.8 Applications of DNA Barcoding in Discovery and Monitoring

DNA barcoding enables the precise identification of species across their life stages and can uncover cryptic diversity. It has been applied in diverse areas, such as food authentication, disease monitoring, and forensics. In biodiversity monitoring, barcoding supports the tracking of community composition and the detection of endangered or invasive species. Barcoding can also illuminate species interactions through analysis of co-occurring DNA, such as microbiomes or parasitic relationships within a host, a concept termed 'symbiome' sequencing. This has implications for ecological modeling and public health surveillance, including the early detection of pandemics. At the population level, barcoding can assess genetic structure and geographic variation, often serving as a cost-effective starting point for more detailed phylogenetic or genomic studies. Metabarcoding of environmental samples, such as water and feces, is beginning to reveal population-level patterns and stress responses in ecosystems. Barcoding is also being integrated into paleoecological studies, providing high-resolution insights into past biodiversity and environments.

2.4.9 Toward Automation and Wider Accessibility in Barcoding

Although DNA barcoding faces challenges, such as limited resolution in some taxa, primer biases, and pseudo-gene amplification, it continues to evolve through technological innovation. Alternative markers (e.g., 12S for fish, trnL for plants) are being used in metabarcoding to address taxon-specific limitations.

New platforms, such as PacBio Sequel, enable extensive multiplexing and cost savings. Tools such as lateral flow assays combined with barcoding primers (e.g., for detecting crown-of-thorns starfish) bring barcoding to field settings with minimal lab infrastructure, making the technique viable for citizen science and conservation in remote areas. Future developments may include fully automated, field-deployable barcoding systems (e.g., FASTFISH-ID), which could provide real-time data on species presence and migration patterns for conservation and biosecurity.

2.4.10 A Global, Inclusive Vision for DNA Barcoding

Global barcoding initiatives, such as iBOL, promote collaboration, with millions of barcode records and a growing user base across over 200 nations. Ensuring fair access to these resources and respecting data sovereignty is essential, particularly in the context of digital sequence information (DSI), which currently falls outside the scope of existing international legal frameworks, such as the Nagoya Protocol. Inclusive and open sharing of barcode data supports both science and equity. Citizen science projects, such as the School Malaise Trap Program, demonstrate how public engagement can enhance data collection, education, and biodiversity awareness.

As long as the barcoding community maintains a focus on data quality and continues to innovate, DNA metabarcoding will remain a central component of biodiversity research. Its scalability, affordability, and expanding range of applications make it a cornerstone method for understanding and preserving global biodiversity. The extent to which these limitations affect ecological interpretation is often dependent on the study's goals.

Through comparative analysis using various OTU clustering thresholds (87–99%) across different environments, researchers found that fungal community structures remained consistent, largely shaped by dominant taxa. Their work also indicated that excluding rare OTUs did not significantly alter community composition, suggesting that the ITS region's shortcomings may have limited impact in many real-world ecological scenarios.

In local studies, the extent of intraspecific ITS variation is often reduced, making species delineation more feasible. In ecological applications, where genus-level identification often suffices for functional interpretation, the need for precise species-level resolution is reduced. However, estimates of species richness can be highly sensitive to data processing methods, as different clustering strategies may yield vastly different OTU counts.

Additionally, primer mismatches and varying amplicon lengths may lead to a biased representation of certain fungal taxa, underscoring the importance of careful consideration during community analysis. The ITS region remains the primary marker for studying fungal communities despite limitations such as inconsistent species-level resolution and PCR artifacts. While alternative genes have been explored, they often pose challenges in primer design due to high sequence variation. Still, ITS remains useful because its multicopy nature improves detection sensitivity.

In ecological studies, particularly those focusing on community patterns rather than exact species identification, ITS performs reliably. Beta diversity analysis, for example, tends to remain stable even when rare species or processing methods vary. Although technical decisions can influence species richness estimates, many studies operate effectively at the genus level, reducing the need for perfect resolution.

As long-read sequencing and metagenomic techniques advance, they offer promising alternatives for comprehensive community profiling. However, these methods face challenges, such as DNA degradation and high costs. For now, ITS remains a practical and accessible tool, especially when paired with expanding reference databases. Ongoing developments in genomics, data curation, and sequencing technologies will likely refine and complement current barcoding approaches.

2.4.11 Advancements and Alternatives to ITS in Fungal Metabarcoding

The long-term viability of the ITS region as a single-locus barcode is uncertain. The development of long-read sequencing and metagenomics now permits the recovery of full-length ribosomal operons. Yet, environmental DNA is often degraded, making amplification of long fragments difficult and potentially skewing community representation due to PCR biases. Metagenomic sequencing offers a theoretically unbiased alternative by sequencing all DNA in a sample; however, the complexity of assembling fungal genomes from environmental samples, especially soils, remains a technical challenge.

Moreover, these approaches require deeper sequencing to detect rare taxa, which can be a costly endeavor. Emerging techniques, such as single-cell genomics, transcriptomics, and microfluidic tools, may eventually enhance or replace some ITS-based methods. Despite these advances, ITS will likely remain valuable for rapid, cost-effective profiling, particularly in studies involving degraded or ancient DNA.

It can also serve as a screening step before more comprehensive genomic analysis. Regardless of the method, accurate species identification depends on high-quality, curated reference databases, whether based on ITS, full rDNA operons, or complete genomes. Such databases, in turn, rely on continued investment in fungal taxonomy and systematics.

2.4.12 Improving Fungal Identification: Toward Reliability and Resolution

To enhance the accuracy of fungal identification, several approaches can be adopted. Specialized resources, such as Q-Bank and the ISHAM ITS barcoding database, offer curated datasets for plant and human/animal pathogens. In addition, third-party annotations integrated with primary repositories, such as GenBank, either directly or through collaborations with curated databases, can improve data reliability.

A broader NCBI RefSeq Targeted Loci dataset, incorporating thoroughly validated sequences, particularly those supported by multilocus phylogenetic studies, would also support more robust identification efforts. One promising approach is a centralized annotation system that links three identifiers: the GenBank accession number, a taxonomic registration number from systems like MycoBank or Index Fungorum, and the DOI of the corresponding validating publication. This structured system could then be embedded into automated workflows.

Interactive identification tools, such as DELTA IntKey, MycoKeys, DiscoverLife, and MyCoPortal, utilize user-selected traits to refine species identification. Tools like PhyloKey integrate morphological and genetic information, generating confidence scores for identifications based on molecular phylogenies. While data-intensive, these methods offer transparent and reproducible outcomes.

2.4.13 Challenges in Achieving Universal Fungal Identification

A major barrier to consistent fungal identification is the prevalence of cryptic species (organisms that are morphologically identical but genetically distinct). Convergent evolution adds complexity, as unrelated species may develop similar traits (Crespo & Pérez-Ortega, 2009; Cai et al., 2011). These issues, widespread across fungal lineages, complicate efforts to develop a single, universal identification system.

Although formal taxonomy may lag due to limited access to molecular tools, recognizing distinct phylogenetic lineages remains essential. Identification strategies must be adapted based on the group studied, the species concept applied, and the tools available. A lack of molecular methods should not excuse inaccurate identifications; mycological expertise can often outperform automated tools in such cases.

In sectors such as medicine, agriculture, and industry, high-resolution identification is critical. DNA barcoding and molecular diagnostics are now standard practices, as phenotypic methods often fall short of providing accurate results. For example, *Candida auris*, a multi-drug-resistant yeast, has frequently been misidentified using conventional techniques. Similarly, identifying plant pathogens like *Phyllosticta citricarpa* requires lineage-specific markers, such as those curated in the Q-Bank system.

In some cases, even strain-level identification is necessary. However, a universally applicable identification method across all fungal groups remains elusive due to the vast diversity and diagnostic needs.

As sequencing technology advances, particularly with the rise of long-read sequencing and metagenomics, the long-term role of ITS as a single-marker approach remains uncertain.

Innovations such as full-length rDNA operon sequencing have enhanced taxonomic resolution; however, the challenges of amplifying long DNA fragments from environmental samples persist, particularly when DNA is degraded.

Additionally, longer targets are more prone to amplification bias, which can distort community composition. Metagenomics provides a promising, less-biased alternative by sequencing all DNA in a sample. However, challenges remain when applying this technique to complex environmental matrices, such as soil, particularly regarding sequencing depth, cost, and consistency across replicates.

Detecting rare organisms via metagenomics also demands high sequencing effort. Emerging techniques, such as single-cell genomics, transcriptomics, and microfluidics, are poised to reshape fungal ecology, potentially replacing ITS barcoding in certain applications. Yet, for cost-effective community screening, particularly when dealing with degraded or ancient DNA, ITS remains a practical choice. It also serves well for preliminary sample screening before more detailed analyses, such as metagenome assembly or transcriptome profiling. Regardless of methodology, the effectiveness of molecular tools is heavily dependent on comprehensive and well-maintained reference databases. Whether using ITS, full operons, or genomic sequences, consistent taxonomic curation is essential and relies on sustained efforts in systematics and taxonomy.

Fungal species, like those of other organisms, are characterized by both horizontal factors, such as phylogenetic relationships and phenotypic traits, and vertical dimensions, including their evolutionary origins and patterns of diversification over time. These distinct evolutionary pathways complicate the application of uniform and distinctive criteria for species delimitation and identification. Effective classification methods must be tailored to specific taxonomic groups, yet uncertainties often remain due to gaps in available identification tools and reference datasets. Additionally, the increasing use of high-throughput techniques, including OTU clustering and BLAST-based similarity searches, tends to magnify these limitations. A comprehensive evaluation of the different conceptual frameworks for delineating fungal species, including those based on reproductive biology, is presently practical only for a limited number of taxa, primarily model organisms. However, insights gained from these models are generally applicable only to closely related or ecologically similar species. Therefore, there is a need to broaden this approach to include representative species across all major fungal groups, encompassing a wide range of phenotypes and evolutionary lineages. For large-scale documentation of fungal diversity, an integrative (polyphasic) taxonomic strategy is the most effective. This approach should be tailored to the specific taxonomic group and incorporate molecular and phenotypic data. In many cases, single-locus DNA barcoding may be sufficient, while more complex taxa may require additional barcoding markers or multilocus analysis (Stielow et al., 2015). For particularly challenging species complexes, phylogenomic methods can provide higher resolution; however, these require substantial computational power and expert input, limiting their use to select case studies at present. Phenotypic characteristics continue to play a vital role in fungal taxonomy, incorporating information from cultures and other relevant sources. In certain taxonomic groups, it remains appropriate for new species to be described based solely on phenotypic evidence, particularly when molecular data are unavailable. Nevertheless, any use of phenotypic traits for species delineation must be supported by thorough and critical analysis. When specimens are suitable for molecular study, but the necessary techniques are inaccessible, collaborative efforts to obtain genetic data are encouraged. The overarching objective is to document all fungal species using molecular tools. Phenotypic data are especially valuable in integrative taxonomy when evaluating the distinctiveness of phylogenetic clades. In such instances, quantitative analysis using structured phenotypic matrices within a phylogenetic framework is essential, as it enables the assessment of trait variation and the identification of reliable diagnostic features. The ITS region continues to be the universal molecular barcode for the initial delineation of fungal phylogenetic lineages, providing a foundational diagnostic marker. In instances where it lacks sufficient resolution to discriminate among closely related species, supplementary barcoding loci or multilocus sequence typing are required to enhance taxonomic resolution and accuracy. The discriminatory power of molecular markers is taxon-dependent, necessitating empirical validation of their efficacy within each fungal group rather than indiscriminate application across taxa. Despite this, ITS is anticipated to remain the principal marker employed in fungal metabarcoding studies. Nonetheless, the incorporation of long-read sequencing technologies or additional barcoding markers has the potential to further refine species-level resolution (Yeates et al., 2011). Moreover, metabarcoding analysis should transition from conventional OTU curation and BLAST-based similarity searches toward phylogenetically informed methodologies, such as read placement through phylogenetic binning, which offer superior taxonomic precision. Current challenges in fungal DNA barcoding stem less from the conceptual framework of the method and more from deficiencies in reference databases. These include incomplete taxonomic representation, insufficient documentation of genetic diversity, and inaccuracies in sequence annotations.

REFERENCES

Adamowicz, S. J. (2015, Fall). *International Barcode of Life Conference 2015: Summary of the 6th International Barcode of Life Conference (iBOL 6)*, Guelph, Canada. Barcode Bulletin.

Agapow, P. M., et al. (2004). The impact of species concept on biodiversity studies. *Quarterly Review of Biology*, 79(2), 161–179.

Aime, M. C., et al. (2018). Pucciniomycotina: Pucciniomycetes. In D. J. McLaughlin & J. W. Spatafora (Eds.), *Systematics and Evolution* (pp. 271–294). Berlin, Germany: Springer.

Ainsworth, G. C., Sparrow, F. K., & Sussman, A. S. (Eds.). (2008). *The Fungi: An Advanced Treatise*. Academic Press.

Alexopoulos, C. J., Mims, C. W., & Blackwell, M. (1996). *Introductory Mycology* (4th ed.). Wiley.

Armaleo, D., & Clerc, P. (1991). Lichen chimeras: DNA analysis suggests that one fungus forms two morphotypes. *Experimental Mycology*, 15(1), 1–10.

Arve, L. E., Gjerde, R., & Johansen, S. D. (2011). Phylogenetic species delimitation within the Fusarium graminearum species complex. *Mycologia*, 103(3), 464–472.

Badali, H., de Hoog, G. S., & Vicente, V. A. (2011). Black yeast fungi: Ecology, pathogenicity, and biotechnological potential. *Fungal Biology Reviews*, 25(1), 1–9.

Becker, M., & Pusharava, T. (2023). The utility of mitochondrial cytochrome c oxidase genes in DNA-based species identification: Advances and applications. *Journal of Molecular Taxonomy*, 12(1), 45–60.

Cai, et al. (2011). Explored phylogenetic relationships within the Sordariaceae family, revealing instances of morphological convergence and cryptic speciation. *Mycol Res*, 110, 137–150.

Covert, S. F., et al. (2007). New fungal species from the Appalachian mountains. *Mycologia*, 99(4), 641–656.

Crespo, & Pérez-Ortega. (2009). Discussed the prevalence of cryptic species in lichens, highlighting the challenges they pose to traditional morphological identification methods. *Mycological Research*, 113(10), 1048–1063.

Crous, P. W., Wingfield, M. J., Burgess, T. I., Hardy, G., & Crane, C. (2009). Fungal identification: Past, present, and future. *Studies in Mycology*, 64, 3–12.

Damm, et al. (2012). Found similar limitations in plant-pathogenic groups within Dothideomycetes and Sordariomycetes, where ITS failed to resolve species-level distinctions. *Studies in Mycology*, 73, 37–113.

Erickson, D. L., Johnson, C. J., & de Hoog, G. S. (2017). Molecular approaches for the identification of human pathogenic fungi: Current status and future prospects. *Medical Mycology*, 55(5), 421–435.

Eyualem, A., & Blaxter, M. (2003). Comparison of biological, molecular, and morphological methods of species identification in nematodes. *Journal of Nematology*, 35(2), 119–128.

Geiser, D. M., Klich, M. A., Frisvad, J. C., Peterson, S. W., Varga, J., & Samson, R. A. (2007). The current status of species recognition and identification in Aspergillus. *Studies in Mycology*, 59, 1–10.

Geiser, D. M., et al. (2013). The current status of species recognition and identification in Aspergillus. *Studies in Mycology*, 75(1), 141–147.

Gilmore, S. R., Gräfenhan, T., Louis-Seize, G., & Seifert, K. A. (2009). Multiple copies of COX1 and COX2 genes in species of the Fusarium sambucinum complex: Implications for molecular species identification. *BMC Microbiology*, 9, 131.

Goulding, T. C., & Dayrat, B. (2016). Integrative taxonomy: Concepts, practices, and implications. *Systematic Biology*, 65(5), 844–851.

Hawksworth, D. L. (2011). One fungus = one name: DNA and fungal nomenclature twenty years after PCR. *IMA Fungus*, 2(2), 113–120.

Hebert, P. D. N., Cywinska, A., Ball, S. L., & deWaard, J. R. (2003). Biological identifications through DNA barcodes. *Proceedings of the Royal Society B: Biological Sciences*, 270(1512), 313–321.

Jiang, et al. (2016–2020). Conversely, certain lichen-forming fungi within Dothideomycetes, such as the genus Strigula, exhibit high resolution with ITS. *MycoKeys*, 68, 1–25.

Kendrick, B. (1979). *The Fifth Kingdom*. Waterloo, Canada: Mycologue Publications.

Kurtzman, C. P., Fell, J. W., & Boekhout, T. (Eds.). (2011). *The Yeasts: A Taxonomic Study* (5th ed.). Elsevier.

Lücking, R. (2019). Fungal taxonomy: Current status and future challenges. *IMA Fungus*, 10, 1–20.

Martin, F. N., & Tooley, P. W. (2003). Phylogenetic relationships among Phytophthora species inferred from mitochondrial cytochrome oxidase I and II genes. *Mycologia*, 95(2), 269–284.

Mattsson, J. E., & Lumbsch, H. T. (1989). Generic and species concepts in lichenized fungi: A case study of the genus Peltigera. *Lichenologist*, 21(1), 1–22.

Messuti, M. I., et al. (2016). Taxonomic studies on the lichen genus Cladonia in southern South America. *Mycologia*, 108(5), 934–949.

Meyer, et al. (2019). Reported that up to 25% of fungi pathogenic to humans or animals could not be reliably identified using the ITS region alone. *Fungal biology and biotechnology*, 7(1), 5.

Moncada, B., et al. (2013). Phylogenetic analysis and taxonomy of the genus Sticta (Ascomycota: Peltigeraceae) in Colombia. *Bryologist*, 116(3), 265–285.

Moritz, C. (1994). Defining 'evolutionarily significant units' for conservation. *Trends in Ecology & Evolution*, 9(10), 373–375.

Nancy, A., & Binnicker, M. J. (2009). Molecular approaches for the detection and identification of fungal pathogens: PCR and beyond. *Clinical Microbiology Reviews*, 22(3), 488–503.

Nilsson, et al. (2008). Analyzed intraspecific ITS variability across fungi, emphasizing the challenges it presents for accurate molecular identification. *Evolutionary bioinformatics*, 4, EBO-S653.

Plitzner, P., Henning, T., Müller, A., Güntsch, A., Berendsohn, W. G., Borsch, T., & Kilian, N. (2019). The Additivity Project: Achieving additivity of structured taxonomic character data by persistently linking them to individual specimens. *Biodiversity Information Science and Standards*, 3, e37178.

Quaedvlieg, W., et al. (2014). Naming the major clades of Dothideomycetes. *Studies in Mycology*, 78, 1–4.

Raja, H. A., Miller, A. N., Pearce, C. J., & Oberlies, N. H. (2017). Fungal identification using molecular tools: A primer for the natural products research community. *Journal of Natural Products*, 80(3), 756–770.

Schoch, et al. (2012). Evaluated multiple DNA regions as potential barcodes for fungi, recommending ITS as a primary marker but acknowledging its limitations in certain taxa. *Integrative Approaches and Secondary Barcoding Markers*. Proceedings of the National Academy of Sciences of the United States of America, 109(16), 6241–6246.

Stielow, et al. (2015). Proposed the use of secondary barcoding markers, such as TEF1α and RPB2, to improve species resolution where ITS falls short. *Persoonia*, 35, 242–263.

Taylor, J. W. (2011). One fungus = one name: DNA and fungal nomenclature twenty years after PCR. *IMA Fungus*, 2(2), 113–120.

Taylor, J. W., Turner, E., Townsend, J. P., Dettman, J. R., & Jacobson, D. J. (2000). Phylogenetic species recognition and species concepts in fungi. *Fungal Genetics and Biology*, 31(1), 21–32.

Taylor, M. J., Freemantle, N., Geddes, J. R., & Bhagwagar, Z. (2006). Early onset of selective serotonin reuptake inhibitor antidepressant action: Systematic review and meta-analysis. *Archives of General Psychiatry*, 63(11), 1217–1223.

Taylor, M. J., McNicholas, C., Nicolay, C., Darzi, A., Bell, D., & Reed, J. E. (2014). Systematic review of the application of the plan–do–study–act method to improve quality in healthcare. *BMJ Quality & Safety*, 23(4), 90–298.

Termorshuizen, A. J., & Arnolds, E. (1997a). The use of morphological and physiological features for fungal species identification. *Mycological Research*, 101(4), 421–428.

Termorshuizen, A. J., & Arnolds, E. (1997b). Morphological limitations in fungal identification: Focus on pathogenic black yeast. *Mycological Research*, 101(4), 421–428.

Vi Hoang, T. T., Hyde, K. D., Doilom, M., Abdel-Aziz, F. A., & Phillips, A. J. L. (2019). DNA barcoding of commercially important mushrooms and their counterfeit products. *Mycology*, 10(3), 175–187.

Vinarski, M. V. (2019). The species problem and its logic: In search of the unity of the concept. *Biology Bulletin Reviews*, 9(3), 196–206.

White, T. J., Bruns, T., Lee, S., & Taylor, J. (1990). Amplification and direct sequencing of fungal ribosomal RNA genes for phylogenetics. In M. A. Innis, D. H. Gelfand, J. J. Sninsky, & T. J. White (Eds.), *PCR Protocols: A guide to Methods and Applications* (pp. 315–322). San Diego, CA: Academic Press.

Wilkins, J. S. (2018). *Species: The Evolution of the Idea* (2nd ed.). Boca Raton, FL: CRC Press.

Wingfield, M. J., Seifert, K. A., & Webber, J. F. (2012). One fungus, one name promotes progressive plant pathology. *Molecular Plant Pathology*, 13(6), 604–613.

Xu, J. (2005). The inheritance of fungal virulence traits through sexual and asexual reproduction. *FEMS Microbiology Letters*, 251(2), 185–192.

Yeates, D. K., et al. (2011). Integrative taxonomy, or iterative taxonomy? *Systematic Entomology*, 36(2), 209–217.

Yurkov, A. M., et al. (2015). Integrative taxonomy reveals hidden diversity in the yeast genus Mrakia. *Fungal Biology*, 119(12), 1133–1144.

3 DNA Barcoding in Insects

Taxonomy and Biodiversity

Muhammad Umair Sial, Rashad Rasool Khan, Jam Nazeer Ahmad, Zain ul Abdin, Muhammad Arshad, and Muhammad Naveed Anwar

3.1 INTRODUCTION

Species identification is a fundamental part of describing and documenting biodiversity. Commonly, identification of species has relied on morphological diagnoses from taxonomic research. Only specialists, such as taxonomists and competent technicians, can effectively identify taxa, as this requires specialized abilities gained through considerable experience (Jinbo et al., 2011). Accurately identifying species has become more crucial as interest in biodiversity has grown across various disciplines, including evolutionary biology, agriculture, ecology, and economics. Nonetheless, there are far fewer taxonomists and other identification specialists today. Therefore, it is necessary to have precise and alternative identification techniques that are accessible to non-experts. Using genetic rather than morphological data for taxonomic identification is one of the most promising methodologies—a concept crucial to numerous biologists (Blaxter, 2003). Simple, rapid, and affordable DNA studies are now possible for researchers studying biodiversity, thanks to advancements in DNA sequencing technologies. Both the taxonomic dilemma and the advancements in biotechnology were significant factors in the development of DNA barcoding.

Hebert et al. (2003) proposed a method to ensure rapid and precise identification of a wide variety of biological specimens by utilizing a primer set to amplify a 648-base pair (bp) region of the mitochondrial cytochrome c oxidase subunit 1 (COI) gene. This method was named 'DNA barcoding.' To advance DNA barcoding as a global standard for eukaryotic sequence-based identification, the Barcode of Life initiative was subsequently launched. The Consortium for the Barcode of Life (CBOL), established in 2004, officially launched this project to create a comprehensive DNA barcode library and a standardized DNA barcoding system. Following this, the International Barcode of Life project (iBOL; International Barcode of Life, 2010) was launched, marking the beginning of a new phase in the Barcode of Life initiative. The typical DNA barcode region, specifically the COI, is highly effective for species identification. For the majority of animal groupings, this region has strong discriminatory power. All animal phyla can utilize this universal primer, which was initially

DOI: 10.1201/9781003538929-3

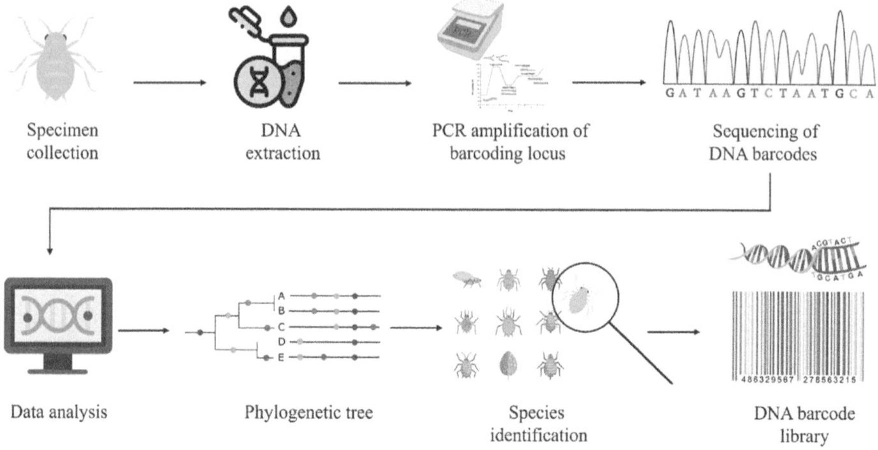

FIGURE 3.1 Illustration of the insect DNA barcoding process from specimen collection to barcode library.

developed for marine invertebrates (Folmer et al., 1994; Hebert et al., 2003, 2004). A sequencer can directly sequence a 648-bp segment since it contains sufficient information. Because this region codes for proteins, alignment is straightforward. By determining whether the obtained sequence is translatable, errors can be identified. The COI region was chosen as the standard DNA barcode due to its beneficial characteristics. For non-experts, DNA barcoding can be a straightforward yet effective technique, particularly for individuals who frequently handle a large number of samples (Figure 3.1).

3.2 PRINCIPLES AND METHODOLOGY OF DNA BARCODING

Based on the covalent ligation of two contiguous oligonucleotide probes when hybridized to a complementary target sequence, DNA barcoding is a type of oligonucleotide ligation assay that is used for SNP detection and genotyping (Raclariu et al., 2018). Barcoding eliminates the need for target pre-amplification, a costly, time-consuming, and error-prone process that can lead to false-positive results when compared with standard nucleic acid detection techniques, such as PCR (Du et al., 2017; Potapov and Ong, 2017). The barcoding process examines variable nucleotides found in standard genomic regions; these nucleotides are important informative characteristics for species classification. For example, the COX I gene is an index used to identify animal species (Bucklin et al., 2011). Fluorescence spectrophotometry can classify barcodes since they are often labeled with spectrally distinct fluorophores. The information is referred to GenBank to create new genetic sequence libraries or to trace the observed samples in phylogenetic categories (Kress, 2017).

Insect DNA barcoding requires several steps, including DNA extraction procedures, PCR amplification, and sequencing techniques. Several DNA extraction techniques have been employed, including commercial kits and Chelex-based extraction

(Janzen et al., 2009). Universal primers designed for the COI gene, or specialized primers for additional gene regions, are used for PCR amplification of target genes (Pentinsaari, 2014). DNA barcode sequences are obtained using high-throughput or Sanger sequencing technologies (Miller, 2013). Existing reference sequences from archives such as GenBank and Barcode of Life Data Systems (BOLD) are compared with the DNA sequences acquired during the sequencing process. For future use and public access, the coded DNA barcode data—which includes specimen details, DNA sequences, and related metadata—is retained in a database or library (Sheth & Thaker, 2017).

3.3 APPLICATIONS IN INSECT TAXONOMY

The database contains several types of model species that have been identified using barcodes. A DNA barcode has been obtained through specific methods, representing a species group for comparison with a model specimen of that species group. Brown et al. (2003) defined a new species category by using a DNA holotype barcode. He also described new types of *Xenothictis* (Lepidoptera: Tortricidae). Following the discovery of a new species group through paratype or DNA holotype barcodes, this method has recognized different phyla, including arthropods (Jinbo et al., 2011). The taxonomic process of categorizing a group of individuals or populations represented as a single species is known as species discovery. The discovery of new species has amplified the value of DNA barcodes. DNA barcoding has certain benefits as it has been used to represent previously unknown and cryptic species (Hebert et al., 2004; Janzen et al., 2005). DNA barcode information is helping to organize the species of related taxa, even with little knowledge about those species (Smith et al., 2008).

DNA barcoding has simplified the identification of new insect species. It also identifies the boundaries of cryptic species groups. For example, DNA barcoding revealed previously unknown species of moths, wasps, and beetles (Pentinsaari, 2014; Miller, 2013; Mitchell, 2008). The method has proven highly beneficial in tropical areas with a rich variety of insects (Janzen et al., 2009). Hidden species have been uncovered through DNA barcoding and cryptic diversity analysis. This analysis is also critical for improving our understanding of evolutionary patterns and processes. It advances our knowledge of biodiversity, supports conservation initiatives, and helps clarify species concepts. It is often impossible to precisely identify insect species due to phenotypic similarity among related and interconnected species (Pigliucci, 2005). Therefore, new DNA barcoding techniques were required to identify all species varieties. DNA barcoding can differentiate cryptic species, which aids in analyzing phylogeny (Almeron et al., 2018). It also explains insect groups, population structures, and their relationships within ecosystems. DNA barcodes from different niches have been used to assess the variation and network of insect communities (Virgilio, 2020). Endangered or threatened species have been monitored efficiently with the help of DNA barcoding to support conservation strategies (Figure 3.2) (Mitchell, 2008).

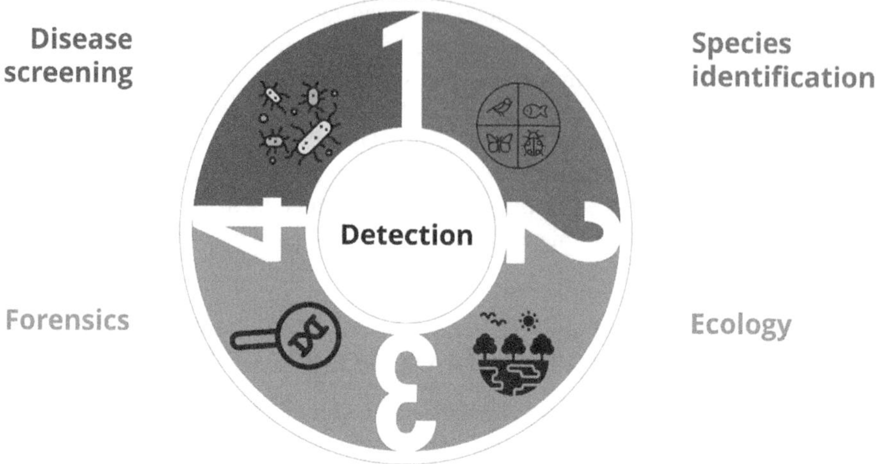

FIGURE 3.2 Applications of DNA barcoding in principal research disciplines.

3.4 DNA BARCODING AND INSECT BIODIVERSITY

The most effective way to combat the decline in insect populations, which harms ecosystems, is likely to implement scalable methods for ongoing biomonitoring and to gain a thorough understanding of insect diversity. It is concerning that we are only now beginning to understand how tropical forests can be rich in species (Srivathsan et al., 2019), how these forests are vertically structured (de Souza Amorim et al., 2022), and how mangroves support a diverse fauna of insects (Yeo et al., 2021). Biodiversity loss is ranked as one of the three global threats by the World Economic Forum for the next ten years. Global patterns have been appearing publicly, as the domination of 20 families of insects in Malaise traps has been revealed worldwide (Srivathsan et al., 2023). These are just a few examples of our limited understanding of insect biodiversity. There is a growing urge to increase the potential for discovering biodiversity and monitoring insect populations. DNA barcoding has been one of the most appealing technologies for this initiative regarding insect biodiversity because robots are used to perform many tasks in daily routines. Of the 15 million insect barcodes in BOLD's database, 12.5 million are publicly accessible; the remaining 2.5 million or so are withheld until the related research is published. BOLD data exhibit comparable geographic coverage to all Global Biodiversity Information Facility data. It would be valuable to see countries allocate funds to fill these knowledge gaps about species diversity. While some countries have been sampled extensively (Canada and Costa Rica account for half of all barcodes, with 2.2 and 5 million, respectively), many others have insufficient data (Meier et al., 2025).

Certain stag beetles, saved as pets, and birdwing butterflies, used as ornaments, are examples of protected or endangered insects that can be illegally traded and detected using DNA barcodes. The early identification of dangerous pests and/or invasive species could prevent their establishment in agriculture and conservation biology. Due to the sudden rise in identification of potential targets, applying different methods to

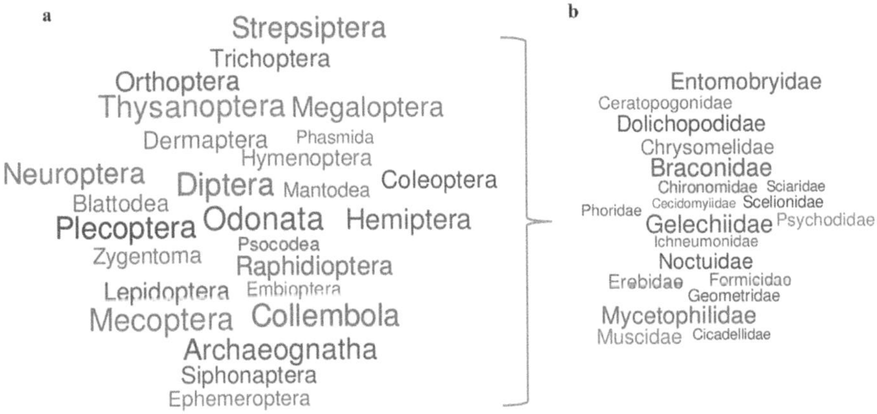

FIGURE 3.3 Word cloud representing the summary of sequenced hexapod (a) orders and (b) families in DNA barcode datasets (Meier et al., 2025).

these targets is essential (Bonants et al., 2010). Rapid identification techniques would facilitate the monitoring of disease vectors, such as mosquitoes, for epidemiological purposes. According to Floyd et al. (2010), DNA barcoding can become a standard method for identifying species in these domains. The biggest threat to biodiversity is invasive pests. Biosecurity relies on the rapid and accurate detection of these threats. Globally, the coverage of the DNA barcoding library is quite valuable for this reason. Armstrong and Ball (2005) evaluated the effectiveness of DNA barcoding-based identification against earlier molecular-based techniques, and they found that DNA barcoding was a superior option for detecting unpredictable species and monitoring pests (Figure 3.3).

3.5 CASE STUDIES IN INSECT DNA BARCODING

Different studies suggest that the accuracy of identifying both pests and beneficial organisms can be improved for Integrated Pest Management (IPM) using DNA barcoding technology (Naaum et al., 2012). The IPM strategies utilize natural features, such as phenology and pesticide susceptibility, to prevent pests from becoming established. For example, identifying two apple pests, *Aphis spiraecola* and *Aphis pomi*, is also tricky because they closely resemble each other (Foottit et al., 2009; Naaum et al., 2012). The use of pesticides is the most common method of controlling these pests, but resistant biotypes develop very quickly (Lowery et al., 2006). There is a difference between *Aphis spiraecola* and *Aphis pomi* in the hibernating host, which affects management strategies and timing. These two insect pests can be differentiated using DNA barcoding, which proves helpful for incorporating them into pest management programs targeting the pest complex (FAO, 2006). It has been found that morphological differences exist in species such as leaf miners, including *Liriomyza* spp. (which consist of over 376 species), which are difficult to distinguish. These species demonstrate variation in their reproductive potential and vulnerability to chemical and biological control (Gao et al., 2011). Researchers believe that these biological differences have contributed to the replacement of *Liriomyza sativae* (Blanchard) by *Liriomyza trifolii* (Burgess) within California and China, and the substitution of *L. trifolii* with *L. sativae* in Japan (Gao et al., 2012). Considering the control measures that recognize the potential for unstable pest complex disturbance, DNA barcoding helps improve management decisions by precisely detecting the four most abundant *Liriomyza* pest species (Scheffer et al., 2006). In IPM, DNA barcoding has the most excellent application when there is non-identical similarity in appearance but differences in life cycle, threshold of damage, and vulnerability to various control methods.

For many organisms, the molecular approach of DNA barcoding simplifies the development of species assumptions, which may improve pest management. For example, two sympatric populations of the western flower thrips, *Frankliniella occidentalis* (Pergande), are suspected to occur in New Zealand and California (Rugman-Jones et al., 2010). This species, *F. occidentalis*, which has 18 recognized synonyms, exhibits morphological and behavioral variation, the causes of which might partly explain this split. In addition, powerful DNA barcode data and mating studies are challenging the species status of *Bemisia tabaci*. Most individuals in *B. tabaci* fit previously recognized biotypes; however, some exhibit differing economically important traits, such as levels of pesticide resistance and the capacity to transmit plant viruses (De Barro et al., 2011). Remarkably, barcoding of DNA has successfully separated the critical biotypes of *B. tabaci, which is* consequential for plant developmental processes. It has also helped to pinpoint a couple of species of the fruit pest *Zaprionus indianus* (Drosophilidae) (Yassin et al., 2008). Using morphological and behavioral data, as well as DNA barcoding, the 'Braconidae' type specimen of the Korean soybean aphid *Binodoxys koreanus*, a cryptic parasitoid species, was identified through a single fixed nucleotide polymorphism. Similarly, barcoding was used to identify a native North American leaf-mining moth, *Antispila oinophylla* (van Nieukerken et al., 2012).

Many analytical approaches exist to study the existence of cryptic species, but these approaches are often used to test existing assumptions. Barcoding helps identify known species and suggests that some may be unidentified, supporting the claim that it can be used to test species hypotheses in an integrative taxonomic approach. DNA barcoding is utilized in insects collected from crime scenes or in association with human remains in forensic casework to help estimate the linkage of crime scenes to suspects, the time elapsed since death, and the place of origin. The DNA barcoding reference database, constructed for forensic purposes and housed at the Bavarian State Collection of Zoology, contains approximately 502 high-quality COI sequences from 88 species of arthropods (Chimeno et al., 2018). DNA barcoding facilitates the work of forensic specialists, which in turn assists police and the justice system in solving crimes by increasing the reliability and accuracy of the forensic work. When a corpse is in an advanced state of decomposition, forensic entomology serves as a vital diagnostic tool. As long as arthropod colonization is not hindered by cold weather or other limitations, a forensic entomologist can determine the postmortem interval or the maximum period between death and burial (Tarone and Sanford, 2017). It is widely accepted that colonization by arthropods occurs at the moment of death. Accurate species identification is crucial in any death investigation involving arthropods (Figure 3.4) (Joseph et al., 2011).

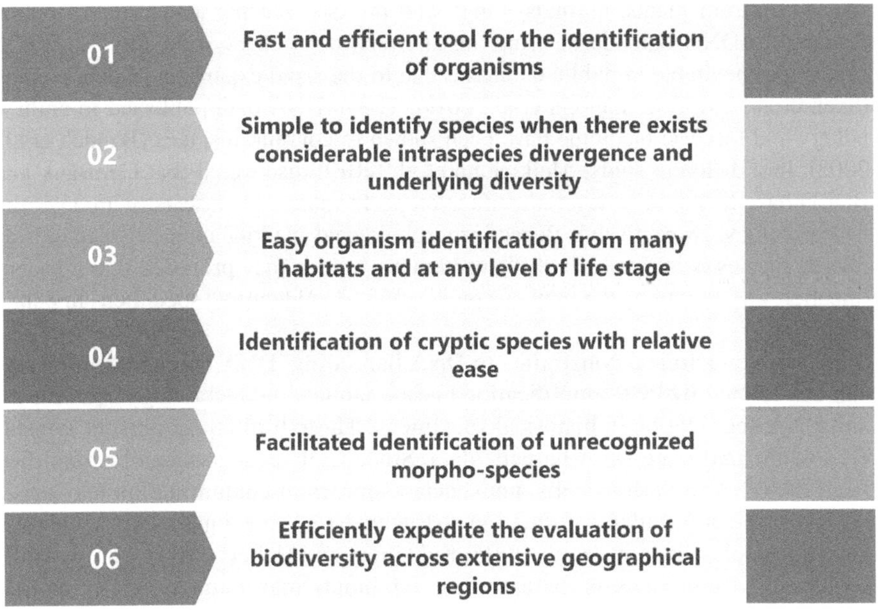

FIGURE 3.4 Major strengths of insect DNA barcodes.

3.6 CHALLENGES AND LIMITATIONS OF DNA BARCODING

Extensive reference collections with DNA barcodes from recognized species are essential for the success of DNA barcoding. Effective species identification is complicated by the absence of comprehensive reference records, particularly for recently discovered or unexplored species. The reliability of species identification using DNA barcodes depends on the quality of the databases. The degree of convergence between intraspecific and interspecific genetic variations is crucial when comparing the query and reference sequences. Misidentifications or other taxonomic problems are likely highlighted by an aberrant intraspecific distance that is greater than the threshold or an interspecific distance that is less than the threshold (Hebert et al., 2003; Gwiazdowski et al., 2015). This type of data is not accurately distinguished; therefore, this method can make it challenging to identify species. Library resources are frequently updated and maintained. Collins and Cruickshank (2013) provided data on one of the problems often encountered in barcoding investigations. They offered solutions for each issue and highlighted the impracticality of using the same classification method for identifying different groups of species, as well as the inapplicability of consistent genetic variation thresholds across various taxonomic groups. The fact that species often exhibit paraphyly in gene trees is one of the main reasons why DNA barcodes may not be able to differentiate previously identified species (Funk and Omland, 2003; Thalmann et al., 2004; Tautz et al., 2003).

Several barcode sequences have been developed and added to open databases like the BOLD and GenBank. BOLD presently has barcode sequences for 11,429,650 specimens from plants, animals, fungi, and protists, serving as a primary public resource of DNA barcode data and an analysis tool (retrieved on July 12, 2022). Errors are inevitable in public databases due to the rapid expansion of data volume (Shen et al., 2013; Meiklejohn et al., 2019). The mtDNA data published in anthropology and forensic medicine have been shown to contain mistakes (Bandelt et al., 2002). In a different study, Meiklejohn et al. (2019) discovered that GenBank and BOLD have low species-level identification performance for insects (with 35% and 53% accuracy, respectively). Because morphological identification between closely related species is intrinsically difficult in many orders, they proposed that misidentification was caused by the previous publication of unidentified specimens in public collections.

There are technical constraints to DNA barcoding. DNA barcoding in insects can be affected by PCR amplification biases, sequencing technology constraints, and DNA degradation in historical specimens. These limitations must be considered while analyzing DNA barcode data. Since COI sequences can be amplified from the DNA of various cells, unexpected sources of contamination may arise; however, the molecular information extracted from specimens can accurately identify species (Ivanova et al., 2006; Kekkonen and Hebert, 2014). For example, sequences of commensals, parasites, or symbionts may unintentionally be discovered when collecting host tissues for DNA extraction. Further contamination may occur if DNA extraction is not performed precisely according to experimental guidelines (Wilson et al., 2019). A dynamic evaluation of morphological characteristics and barcodes would be ideal as the foundation for the final species identification results in a DNA barcoding study. Nevertheless, this procedure is sometimes overlooked. The dependability of downstream analysis and future data reuse may be jeopardized if mistakes are made during the production of sequence data and public databases, which could have an adverse downstream effect (Barratt et al., 2018; Peres et al., 2021).

3.7 FUTURE DIRECTIONS AND INNOVATIONS

Barcoding of DNA is one of the most widely accepted and standardized tools for species identification. Sanger sequencing is a significant component of traditional DNA barcoding; however, next-generation sequencing (NGS) technologies can enhance and complement it (Batovska et al., 2017; Hebert et al., 2018). NGS can be a valuable alternative to Sanger sequencing in certain DNA barcoding research applications where a large number of specimens need to be analyzed. Co-amplified products and other PCR products can be sequenced concurrently with it. Using these new scientific instruments, one can provide additional data for fresh and museum insect specimens. DNA barcoding can be used in conjunction with morphology, natural history, and other information to identify and describe insect specimens, although it may not be an adequate method for describing species independently. Enabling taxonomists to sort specimens quickly and identify divergent taxa that may belong to new species will speed up the rate of species identification (Hebert and Gregory, 2005). Multiple nations have demonstrated that highly targeted, group-specific barcode collections can aid in the faster generation of DNA barcode data (Huemer et al., 2020).

Although sequencing has significantly decreased costs and processing time, barcoding and metabarcoding are unlikely to become the sole methods for economically viable biomonitoring. Artificial intelligence (AI) techniques for analyzing samples based on pictures (Valan et al., 2019; Wührl et al., 2022; Fujisawa et al., 2022) are anticipated to be crucial for quick and affordable identification in the future. The pervasive link between abundance and diversity of species aids in developing these tools. In biology, bulk samples consistently show that a small number of species account for a significant percentage of the specimens (Geiger et al., 2016). DNA-based identification may focus on species for which AI algorithms have not yet been developed if these species cannot be recognized from photos. This implies that imaging should be a key consideration for DNA barcoding projects.

REFERENCES

Almeron SF, Sperb C, Castilho CL, Figueiredo PCC, Goncalves LT and Machado R (2018). Molecular identification of shark meat from local markets in southern Brazil based on DNA barcoding: Evidence for mislabeling and trade of endangered species. *Front. Genet.* 9(1), 138.

Armstrong KF and Ball SL (2005). DNA barcodes for biosecurity: Invasive species identification. *Philosophical Transactions of the Royal Society of London Series B* 360, 1813–1823.

Bandelt HJ, Quintana-Murci L, Salas A and Macaulay V (2002). The fingerprint of phantom mutations in mitochondrial DNA data. *Am. J. Hum. Genet.* 71, 1150–1160.

Barratt BIP, Cock MJW and Oberprieler RG (2018). Weevils as targets for biological control, and the importance of taxonomy and phylogeny for efficacy and biosafety. *Divers. Basel* 10, 73.

Batovska J, Cogan NOI, Lynch SE and Blacket MJ (2017). Using next-generation sequencing for DNA barcoding: Capturing allelic variation in ITS2. G3-. *Genes Genom. Genet.* 7, 19–29.

Blaxter M (2003). Counting angels with DNA. *Nature* 421, 122–124.

Bonants P, Groenewald E, Rasplus JY et al. (2010). QBOL: A new EU project focusing on DNA barcoding of quarantine organisms. *EPPO Bulletin* 40, 30–33.

Brown JW, Miller SE and Horak M (2003). Studies on New Guinea moths. 2. Description of a new species of *Xenothictis* Meyrick (Lepidoptera: Tortricidae: Archipini). *Proceedings of the Entomological Society of Washington* 105, 1043–1050.

Bucklin A, Steinke D and Blanco-Bercial L (2011). DNA barcoding of marine metazoa. *Ann. Rev. Mar. Sci.* 3, 471–508.

Chimeno C, Moriniere J, Podhorna J, Hardulak L, Hausmann A and Haszprunar G (2018). DNA barcoding in forensic entomology-establishing a DNA reference library of potentially forensic relevant Arthropod species. *Journal of Forensic Sciences* 64(2), 593–601.

Collins RA and Cruickshank RH (2013). The seven deadly sins of DNA barcoding. *Mol. Ecol. Res.* 13(6), 969–975.

De Barro PJ, Liu SS, Boykin LM and Dinsdale AB (2011). Bemisia tabaci: A statement of species status. *Annual Review of Entomology* 56, 1–19.

de Souza Amorim D, Brown BV, Boscolo D et al. (2022). Vertical stratification of insect abundance and species richness in an Amazonian tropical forest. *Sci. Rep.* 12, 1734.

Du K, Park M, Griffiths A, Carrion R, Patterson J, Schmidt H and Mathies R (2017). Microfluidic system for detection of viral RNA in blood using a barcode fluorescence reporter and a photocleavable capture probe. *Anal. Chem.* 89, 12433–12440.

FAO (2006). ISPM No. 27 Diagnostic Protocols for Regulated Pests. International Standards for Phytosanitary Measures. Rome, IPPC.

Floyd R, Lima J, deWaard J, Humble L and Hanner R (2010). Common goals: Policy implications of DNA barcoding as a protocol for identification of arthropod pests. *Biological Invasions* 12, 2947–2954.

Folmer O, Black M, Hoeh W, Lutz R, Vrijenhoek R (1994). DNA primers for amplification of mitochondrial cytochrome c oxidase subunit I from diverse metazoan invertebrates. *Molecular Marine Biology and Biotechnology* 3, 294–299.

Foottit RG, Lowery DT, Maw HEL, Smirle MJ and Lushai G (2009). Identification, distribution, and molecular characterization of the apple Aphids Aphis pomi and Aphis spiraecola (Hemiptera: Aphididae: Aphidinae). *Canadian Entomologist* 141, 478–495.

Fujisawa T, Noguerales V, Meramveliotakis E, Papadopoulou A and Vogler AP (2022). Image-based taxonomic classification of bulk insect biodiversity samples using deep learning and domain adaptation. *Syst. Entomol.* 48, 387–401.

Funk DJ and Omland KE (2003). Species-level paraphyly and polyphyly: Frequency, causes, and consequences, with insights from animal mitochondrial DNA. *Annu. Rev. Ecol. Evol. Syst.* 397–423.

Gao YL, Lei ZR, Abe Y and Reitz SR (2011). Species displacements are common to two invasive species of leafminer fly in China, Japan, and the United States. *Journal of Economic Entomology* 104, 1771–1773.

Gao YL, Reitz SR, Wei QB, Yu WY and Lei ZR (2012). Insecticide-mediated apparent displacement between two invasive species of leafminer fly. *PLoS One* 7, e36622.

Geiger MF, Moriniere J, Hausmann A et al. (2016). Testing the global Malaise trap program—how well does the current barcode reference library identify flying insects in Germany? *Biodivers. Data J.* 4, e10671.

Gwiazdowski RA, Foottit RG, Maw HEL and Hebert PDN (2015). The Hemiptera (insecta) of Canada: Constructing a reference library of DNA barcodes. *PLoS One* 10, e0125635.

Hebert PDN, Braukmann TWA, Prosser SWJ, Ratnasingham S, DeWaard JR, Ivanova NV, Janzen DH, Hallwachs W, Naik S, Sones JE and Zakharov EV (2018). A sequel to Sanger: Amplicon sequencing that scales. *BMC Genomics* 19, 219.

Hebert PDN, Cywinska A, Ball SL and deWaard JR (2003). Biological identifications through DNA barcodes. *Proceedings of the Royal Society of London, Series B* 270, 313–321.

Hebert PDN and Gregory TR (2005). The promise of DNA barcoding for taxonomy. *Syst. Biol.* 54(5), 852–859.

Hebert PDN, Penton EH, Burns JM, Janzen DH and Hallwachs W (2004). Ten species in one: DNA barcoding reveals cryptic species in the neotropical skipper butterfly *Astraptes fulgerator*. *Proceedings of the National Academy of Sciences of the United States of America* 101, 14812–14817.

Huemer P, Karsholt O, Aarvik L, Berggren K, Bidzilya O, Junnilainen J, Landry JF, Mutanen M, Nuppone K, Segerer A and Šumpich J (2020). DNA barcode library for European Gelechiidae (Lepidoptera) suggests greatly underestimated species diversity. *ZooKeys* 921, 141.

International Barcode of Life (2010). *International Barcode of Life* [Cited 15 October 2010]. http://ibol.org/.

Ivanova NV, deWaard JR and Hebert PDN (2006). An inexpensive, automation-friendly protocol for recovering high-quality DNA. *Mol. Ecol. Notes* 6, 998–1002.

Janzen DH, Hajibabaei M, Burns JM, Hallwachs W, Remigio E and Hebert PDN (2005). Wedding biodiversity inventory of a large and complex Lepidoptera fauna with DNA barcoding. *Philosophical Transactions of the Royal Society of London, Series B* 360, 1835–1845.

Janzen DH, Hallwachs W, Bladin P, Burns JM, Chacon I and Dapkey T (2009). Integration of DNA barcoding into an ongoing inventory of complex tropical biodiversity. *Molecular Ecology Resources* 9(S1), 1–26.

Jinbo U, Kato T and Ito M (2011). Current progress in DNA barcoding and future implications for entomology. *Entomological Science* 14(2), 107–124.

Joseph I, Mathew DG, Sathyan P and Vargheese G (2011). The use of insects in forensic investigations: An overview on the scope of forensic entomology. *Journal of Forensic and Dental Sciences* 3(2), 89–91.

Kekkonen M and Hebert PDN (2014). DNA barcode-based delineation of putative species: Efficient start for taxonomic workflows. *Mol. Ecol. Resour.* 14, 706–715.

Kress WJ (2017). Plant DNA barcodes: Applications today and in the future. *J. Syst. Evol.* 55, 291–307.

Lowery DT, Smirle MJ, Foottit RG and Beers EH (2006). Susceptibilities of apple Aphid and Spirea Aphid collected from apple in the Pacific Northwest to selected insecticides. *Journal of Economic Entomology* 99, 1369–1374.

Meier R, Lawniczak MK and Srivathsan A (2025). Illuminating entomological dark matter with DNA barcodes in an era of insect decline, deep learning, and genomics. *Annual Review of Entomology* 70(1), 185–204.

Meiklejohn KA, Damaso N and Robertson JM (2019). Assessment of BOLD and GenBank–their accuracy and reliability for the identification of biological materials. *PLoS One* 14, e0217084.

Miller SE (2013). DNA barcoding and the taxonomy of Microgastrinae wasps (Hymenoptera, Braconidae): Impacts after 8 years and nearly 20,000 sequences. *Molecular Ecology Resources* 13(2), 154–164.

Mitchell A (2008). DNA barcoding demystified. *Australian Journal of Entomology* 47(3), 169–173.

Naaum AM, Foottit RG, Maw HEL and Hanner R (2012). Differentiation between Aphis pomi and Aphis spiraecola using multiplex real-time PCR based on DNA barcode sequences. *Journal of Applied Entomology* 136, 704–710.

Pentinsaari M (2014). Barcoding beetles: A regional survey of 1872 species reveals high identification success and unusually deep interspecific divergences. *PLoS One* 9(9), e108651.

Peres PHF, Grotta-Neto F, Luduvério DJ, de Oliveira ML and Duarte JMB (2021). Implications of unreliable species identification methods for Neotropical deer conservation planning. *Perspect. Ecol. Conser.* 19, 435–442.

Pigliucci M (2005). Evolution of phenotypic plasticity: Where are we going now? *Trends in Ecology and Evolution* 20(9), 481–486.

Potapov V and Ong JL (2017). Examining sources of error in PCR by single-molecule sequencing. *PLoS One* 12, e0169774.

Raclariu AC, Heinrich M, Ichim MC and de Boer H (2018). Benefits and limitations of DNA barcoding and metabarcoding in herbal product authentication. *Phytochem. Anal.* 29, 123–128.

Rugman-Jones PF, Hoddle MS and Stouthamer R (2010). Nuclear-mitochondrial barcoding exposes the global pest western flower thrips (Thysanoptera: Thripidae) as two sympatric cryptic species in its native California. *Journal of Economic Entomology* 103, 877–886.

Scheffer SJ, Lewis ML and Joshi RC (2006). DNA barcoding applied to invasive leafminers (Diptera: Agromyzidae) in the Philippines. *Annals of the Entomological Society of America* 99, 204–210.

Shen YY, Chen X and Murphy RW (2013). Assessing DNA barcoding as a tool for species identification and data quality control. *PLoS One* 8, e57125.

Sheth BP and Thaker VS (2017). DNA barcoding and traditional taxonomy: An integrated approach for biodiversity conservation. *Genome* 60(7), 618–628.

Smith MA, Rodriguez JJ, Whitfield JB et al. (2008). Extreme diversity of tropical parasitoid wasps exposed by iterative integration of natural history, DNA barcoding, morphology, and collections. *Proceedings of the National Academy of Sciences of the United States of America* 105, 12359–12364.

Srivathsan A, Ang Y, Heraty JM et al. (2023). Convergence of dominance and neglect in flying insect diversity. *Nat. Ecol. Evol.* 7, 1012–1021

Srivathsan A, Hartop E, Puniamoorthy J et al. (2019). Rapid, large-scale species discovery in hyperdiverse taxa using 1D MinION sequencing. *BMC Biol.* 17(1), 96.

Tarone AM and Sanford MR (2017). Is PMI the hypothesis or the null hypothesis? *Journal of Medical Entomology* 54(5), 1109–1115.

Tautz D, Arctander P, Minelli A, Thomas RH and Vogler AP (2003). A plea for DNA taxonomy. *Trends. Ecol. Evol.* 18(2), 70–74.

Thalmann O, Hebler J, Poinar HN, Pääbo S and Vigilant L (2004). Unreliable mtDNA data due to nuclear insertions: A cautionary tale from analysis of humans and other great apes. *Mol. Ecol.* 13(2), 321–335.

Valan M, Makonyi K, Maki A, Vondracek D and Ronquist F (2019). Automated taxonomic identification of insects with expert-level accuracy using effective feature transfer from convolutional networks. *Syst. Biol.* 68(6), 876–895.

van Nieukerken E, Wagner D, Baldessari M, Mazzon L, Angeli G, Girolami V, Duso C and Doorenweerd C (2012). Antispila oinophylla new species (Lepidoptera, Heliozelidae), a new North American grapevine leafminer invading Italian vineyards: Taxonomy, DNA barcodes and life cycle. *ZooKeys* 170, 29–77.

Virgilio M (2020). DNA barcoding reveals diversity patterns of insect pest species in Southeast Asia. *Journal of Pest Science* 93(1), 997–1010.

Wilson JJ, Sing KW and Jaturas N (2019). DNA barcoding: Bioinformatics workflows for beginners. In *Encyclopedia of Bioinformatics and Computational Biology*, eds. S Ranganathan, K Nakai and C Sconbach (Oxford: Academic Press), 985–995.

Wührl L, Pylatiuk C, Giersch M et al. (2022). DiversityScanner: Robotic handling of small invertebrates with machine learning methods. *Mol. Ecol. Resour.* 22(4), 1626–1638.

Yassin A, Capy P, Madi-Ravazzi L, Ogereau D and David JR (2008). DNA barcode discovers two cryptic species and two geographical radiations In the invasive Drosophilid Zaprionus indianus. *Molecular Ecology Resources* 8, 491–501.

Yeo D, Srivathsan A, Puniamoorthy J et al. (2021). Mangroves are an overlooked hotspot of insect diversity despite low plant diversity. *BMC Biol.* 19(1), 202.

4 DNA Barcoding of Viruses

Tuaseef Tariq, Nabeeha Aslam Khan, Imran Ul Haq, Saira Azmat, Muhammad Waris, Ahmad Nisar, and Hayssam Mohamed Ali

4.1 INTRODUCTION TO DNA BARCODING

In DNA barcoding, a short, standardized piece of the genome is utilized to identify and differentiate among species. It is used extensively in bacteria, plants, and animals, but due to their small, highly variable genomes and rapid rates of evolution, viruses, especially DNA viruses, have a more complex application (Lefeuvre et al., 2019). Notwithstanding these difficulties, DNA barcoding has emerged as a potent instrument in the fields of epidemiology, diagnostics, viral taxonomy, and biodiversity assessment.

The sum of virus particles has been estimated to be more than 10 times larger than the count of cells. The molecular diversity of viruses is also complex, given that their genomes are complex (Aylward and Moniruzzaman, 2022). The classification of viruses and the molecular nature of viruses remain a point of controversy among scientists. DNA barcoding should focus on identifying and explaining the molecular entities of viruses as the key concern (Purty and Chatterjee, 2016). Few studies have been conducted to establish pathogenically significant viruses. Hence, scientists might seek to invent a barcode for identifying or detecting viruses. The identification and classification of viruses present substantial challenges in the fields of virology, epidemiology, and molecular biology due to their immense diversity, rapid mutation rates, and limited morphological features. Traditional virus classification based on morphology, host range, and serology has proven insufficient in the face of rapid viral evolution and the increasing detection of novel and recombinant viral strains (Katsarou et al., 2019). Viruses are both pathogenic and non-pathogenic, responsible for many diseases that affect humans, animals, and plants. Therefore, finding accurate and effective methods to identify them is crucial. DNA barcoding and other molecular techniques have become key methods for detecting viruses, identifying them, and conducting taxonomic studies (Antil et al., 2023).

DNA barcoding is a molecular technique used for species identification and classification that utilizes a standardized, short region of the genome. First proposed by Hebert et al. (2010), this concept refers to the use of a standardized short sequence of DNA from a specific locus of the genome to identify species across the tree of life. Initially designed for use with different animals that utilize the mitochondrial cytochrome c oxidase I (COI) gene, the concept has also been expanded to include

DOI: 10.1201/9781003538929-4

plants, fungi, and microorganisms. In viruses, barcoding presents challenges due to the absence of universally conserved genetic markers and the high mutation rates associated with viral genomes (Beerenwinkel et al., 2012). Barcoding for viral identification has made significant progress in recent years, thanks to advancements in sequencing techniques and bioinformatics tools.

DNA barcoding of viruses primarily serves to improve the classification of viruses, facilitate the rapid and accurate detection of viruses, and enhance the monitoring of different viral groups (Maina and Jones, 2023). Unlike phenotypic identification, which relies on morphology, molecular barcoding is an accurate, continuous, and expanding method for identifying related viruses and strains.

The use of DNA barcoding in viral taxonomy aligns well with the framework provided by the International Committee on Taxonomy of Viruses (ICTV), which classifies viruses based on their genome, replication process, and structure (Siddell et al., 2023). With viral taxonomy shifting towards genes and DNA, DNA barcoding offers a shared and dependable method for distinguishing viral types, particularly when other data, such as morphology or serology, are inconclusive or lacking. As worldwide initiatives like the Global Virome Project (GVP) and One Health work towards better management of viral diversity, DNA barcode technology proves vital for detecting emerging viruses in animals, the environment, and the clinic (Carroll et al., 2018).

In addition to its applications in human health, DNA barcoding is also useful in other fields. In agriculture, DNA viruses that infect plants, such as geminiviruses, are a major cause of crop loss worldwide (Tatineni and Hein, 2023). Through DNA barcoding, we can quickly detect and separate viruses, enabling more effective management plans. DNA barcoding is used in veterinary science to identify infections such as porcine circovirus or bovine herpesvirus (BHV-1), thereby improving disease control and zoonotic prevention (Manessis et al., 2022).

The recent spread of the monkeypox virus, African swine fever virus, and several types of plant-infecting geminiviruses highlights the need for rapid and accurate virus identification. DNA barcoding is considered valuable for determining the origin of new viruses, understanding their spread, tracking their evolutionary changes, and controlling their transmission (Fitzmeyer et al., 2023).

4.1.1 DNA Barcoding of Viruses

The use of unique DNA sequences in barcoding enables the recognition and tracing of DNA viruses within millions of DNA sequences. Using this method allows us to identify pathogens, track virus mutations, and investigate how viruses behave in multiple regions (Pybus and Rambaut, 2009). Barcoding DNA viruses can be achieved and improved by applying different techniques. Using standardized markers at the DNA level, DNA barcoding enables the accurate identification and differentiation of many other species (Antil et al., 2023).

A barcoded virus is designed to exhibit a specific sequence (for example, a barcode) that distinguishes it from other viruses that appear identical in a genomic analysis. The way viruses are barcoded can vary depending on the research, although these viruses generally receive only coding-neutral alterations in their genome (Fitzmeyer et al., 2023). Barcodes often contain between 9 and more than

40 nucleotides. They may be placed in regions that are not translated (e.g., the 3′ or 5′ untranslated regions) or in areas with a few consecutive codons with the same third position (e.g., Leu, Val, Ser, Pro, Thr, and Ala). To add barcodes to the viral genome, custom synthetic gene fragments or site-directed mutagenesis (SDM) are employed. Initially, viruses marked with barcodes were clones, and their barcoded parts were introduced to the viruses through methods such as SDM (Fitzmeyer, 2024). Developing every virus one by one makes it challenging to achieve the same level of diversity found in virus barcodes. Current barcoding methods incorporate degenerate nucleotides during the process, allowing all possible mutations to occur at that location. After amplification and rescuing the barcoded virus, all these mutations can theoretically result in several million unique barcodes, which can be easily generated without going through every barcoded clone (Lauring and Andino, 2011). This technology can only be used for viruses for which scientists already have efficient reverse genetics, cloning, and rescue systems in place.

With DNA barcoding, scientists can identify and observe specific viral lineages within large groups of viruses. Initially, the DNA barcoding method was employed in ecology to aid in species identification. It now involves including short, special DNA sequences, also known as barcodes, inside viral genomes. Genetic tags, called barcodes, help researchers distinguish between strains when the same virus infects multiple people (Blundell and Levy, 2014). Using this technique, experts can now examine the behavior of viruses. It is now possible to explore the details of virus distribution, the steps of transmission, the paths of evolution, and the relationships between viruses and their hosts.

Viral barcoding involves the incorporation of short nucleotide sequences into viral genomes to facilitate the tracking and monitoring of viruses in diseases, their evolution, and in medicine (Jeske, 2018). Traditional barcoding in animals and plants utilizes well-conserved genes, such as COI. However, in viruses, the fast mutation rate and diverse genome types (RNA/DNA, single/double strands) pose additional challenges. There are two main methods: (i) Some genomes of viruses, such as HIV and HCV, contain naturally highlighted regions known as natural barcodes. The accumulation of mutations in such areas makes it simple for scientists to identify the origin of viruses or outbreaks. (ii) Building barcodes: Short random sequences (about 12–20 bp), called barcodes, are added to regions that do not affect the virus's replication in the viral genome. By engineered barcodes, researchers can study the tracking of virus movement, evaluate the effectiveness of vaccines, and choose the best gene therapy approaches (Ghosh et al., 2020).

4.1.2 Molecular Markers for DNA Barcoding of Viruses

DNA barcoding is used to identify species, examine biodiversity, and carry out phylogenetic studies by using well-standardized genetic markers. Although mitochondrial and chloroplast genes are effective for animals, plants, and fungi, viruses—lacking cell structure and possessing varied genomes (DNA or RNA, sometimes single-stranded)—must be identified using additional molecular markers to ensure accuracy. Due to their high mutation rates, compact genomes, rapid evolution, genetic diversity, and the lack of universal marker genes, viral barcoding often targets highly conserved yet variable regions within essential genes, allowing for differentiation at the species or strain level. The most commonly used viral barcoding markers include polymerase genes (RdRp for RNA viruses, pol for retroviruses), structural protein genes (capsid, envelope), and other conserved genomic regions (coat protein in plant viruses, hemagglutinin (HA) in influenza).

For RNA viruses, the gene encoding RNA-dependent RNA polymerase (RdRp) serves as a primary barcoding marker due to its essential role in viral replication and moderate conservation, providing a balance between variability for strain differentiation and stability for broad detection. This marker is utilized in RNA viruses, such as poliovirus of the Picornaviridae, SARS-CoV-2 in the Coronaviridae, and Flaviviridae viruses like dengue and Zika (Babaian and Edgar, 2022a). In retroviruses, the reverse transcriptase (RT) gene within the pol region is often targeted due to its functional conservation and usefulness in studying drug resistance, especially in HIV. The hemagglutinin (HA) gene is also an important marker for RNA viruses, as it enables the subtype classification of influenza viruses (Orthomyxoviridae) into H1N1 and H5N1 and is crucial for host cell entry.

DNA viruses utilize the DNA polymerase (DNA pol) gene as a common barcode, particularly for larger viruses such as Herpesviridae (HSV-1, EBV) and Poxviridae (smallpox and vaccine). Additionally, the MCP (major capsid protein) gene is used for classifying viruses in the Papillomaviridae (HPV) and Polyomaviridae families, where capsid proteins are crucial for interaction with the host and exhibit sufficient genetic variation to differentiate between strains. In plant viruses, barcoding is primarily performed using the coat protein (CP) gene, particularly among tobamoviruses and potyviruses, due to their importance in virion structure and specificity to hosts (Jeske, 2018).

Together with single-gene barcodes, using a combination of different DNA segments is now common to achieve better resolution for viruses that are diverse or undergo frequent recombination. Gag, pol, and env regions are commonly used together in HIV genotyping to enhance the accuracy of the process. In HCV, the initial screening is done with the 5′ untranslated region (5′ UTR), and the NS5B or E1/E2 regions are used to pick out fine differences between strains (Lewitus et al., 2024). The addition of metagenomic sequencing to viral barcoding has enabled the identification of novel viruses, particularly in unknown or environmental samples, as traditional PCR does not apply to these.

Although these types of markers are useful, challenges still exist, such as the high rate of changes, recombination between groups, and problems caused by primers not matching in variable areas. To address these issues, consensus degenerate primers (e.g., those used in pan-viral PCR) and next-generation sequencing (NGS) have improved detection breadth. Additionally, bioinformatics tools (e.g., BLAST, phylogenetic clustering) are crucial for analyzing barcode sequences against reference databases, such as GenBank and specialized viral repositories (ICTV, NCBI Virus) (Epstein and Anthony, 2017). Table 4.1 summarizes examples of viral families and their corresponding DNA barcoding target genes and functions.

TABLE 4.1
Examples of Common Barcoding Targets.

Viral Family	Target Gene(s)	Function
Geminiviridae	Rep (replication-associated), CP (coat protein)	Replication and encapsidation
Herpesviridae	UL27 (glycoprotein B), UL55	Entry and replication
Papillomaviridae	L1 (major capsid protein)	Structural protein
Polyomaviridae	VP1, Large T antigen	Capsid and regulatory roles
Poxviridae	DNA polymerase, RNA polymerase subunits	Replication

4.2 DNA BARCODING TECHNIQUES IN DNA VIRUSES

4.2.1 Selection of Barcoding Regions

DNA viruses lack universal genes that are conserved across all families, in contrast to eukaryotes, which frequently use standardized sections for DNA barcoding, such as the mitochondrial cytochrome c oxidase I (COI) gene. The development of a universal viral barcode is made difficult by this innate genetic variability. As a result, the barcoding regions of DNA viruses are chosen based on conserved genes within a particular viral family or genus.

There are two criteria for barcoding regions when choosing a region:

- Conservation across taxa: The gene region to be used must be sufficiently conserved in the taxonomic group of choice to produce reliable, broadly amplifiable primers.
- Sequence variability: There must be enough genetic variability within the area to allow differentiation of isolates, strains, or species within the group.

Family specificity or genus specificity of detection and identification is ensured by this family-specific or genus-specific method, which plays a crucial role in basic studies of viroids, their taxonomy, evolutionary studies, and diagnostics.

4.2.2 Techniques Used

4.2.2.1 PCR-Based Barcoding

PCR-based barcoding is the most popular and basic method for identifying and classifying organisms, including viruses, based on their genetic composition. This procedure involves designing specific primers against areas of conservation in the genome or regions that are highly conserved among multiple species or strains within a genus or family. These conserved regions provide a stable environment for primer binding during PCR, allowing the target DNA fragment to be amplified (Iqbal et al., 2013). The DNA obtained after amplification is cleaned and sequenced. Thereafter, these sequences are compared to entries already available in open-access reference databases, such as GenBank or the Barcode of Life Data Systems (BOLD). This comparison allows the researcher to identify the organism or assess its evolutionary relationships. The choice of conserved gene regions in DNA viruses is group-specific because viruses lack universal genes like those found in other organisms in the biosphere. This technique is highly sensitive, inexpensive to conduct, and suitable for both environmental and clinical samples, making it a very useful method in molecular taxonomy, biodiversity research, and viral diagnostics (Pereira et al., 2008).

4.2.2.2 Real-Time PCR (qPCR) in DNA Barcoding and Viral Diagnostics

In real-time PCR (quantitative PCR (QPCR) or qPCR), an enhanced version of conventional PCR, a specific DNA fragment is not only amplified, but the quantity of copies is also measured in real time. In the context of DNA barcoding, qPCR enhances identification techniques by introducing a second quantitative dimension of information, making it especially useful in diagnostic applications and viral load screening. This method is based on fluorescently labeled probes or dyes that give a signal corresponding to the quantity of amplified DNA, allowing the entire PCR amplification process to be continuously monitored (Singh and Roy-Chowdhuri, 2016). The primary feature of qPCR is its ability to detect and quantify the presence of viral DNA with high sensitivity and specificity. This makes it a potent tool for detecting DNA viruses in clinical or environmental samples and for tracking the infection process, including approximating the viral load in infected tissues or bodily fluids (Ginzinger, 2002). Unlike traditional PCR, which involves an additional procedure after amplification such as gel electrophoresis, qPCR allows rapid, computerized, and contamination-minimizing workflows. Beyond diagnostics, qPCR is significant in epidemiological studies, vaccine efficacy research, and pathogen surveillance, where accurate quantification of viral presence is essential (Jyothy et al., 2020).

4.2.2.3 Loop-Mediated Isothermal Amplification (LAMP) for DNA Barcoding and Viral Identification

The LAMP method is a highly sensitive, low-cost, and specific DNA amplification technique that works under isothermal conditions, normally 60–65 °C. In contrast to PCR, which requires thermal cycling, LAMP can be performed using simple but effective equipment, such as a water bath or heating block; thus, it is well-suited for field work and point-of-care diagnostics. The implementation process is particularly useful in resource-limited environments where DNA virus detection is required, as it is both fast and simple to perform (Silva et al., 2019). LAMP is a highly specific and efficient method, where each reaction employs four to six primers that specifically recognize and bind to overlapping target DNA sequences. Barcoded primers may also be included in the LAMP reaction to facilitate the concurrent identification and detection of multiple viral targets in the same experiment, a capability known as multiplex LAMP. The outcomes can be observed by the naked eye (turbidity, fluorescence, or colorimetric) and do not require sophisticated detection systems (Padzil et al., 2021). LAMP is increasingly used for the real-time detection of viral pathogens in plant virology, clinical diagnostics, and environmental monitoring due to its simple, robust performance and short turnaround time (typically less than 30 minutes).

The fact that LAMP can operate in decentralized environments without compromising accuracy makes it a valuable addition to conventional PCR-based barcoding strategies (Becherer et al., 2020).

4.2.2.4 Next-Generation Sequencing for DNA Barcoding and Viral Genomics

Another revolutionary development in molecular diagnostics and DNA barcoding is next-generation sequencing (NGS), which enables complete sequencing of genomes or targeted gene sequences in reduced timeframes and with greater depth. Compared to traditional PCR-based methods, which require prior knowledge of target sequences, NGS allows untargeted or semi-targeted analysis, suitable for identifying both known and novel DNA viruses. NGS can generate enormous amounts of information from a single sample, providing the ability to analyze viral populations in great detail through high-throughput parallel sequencing (Barzon et al., 2011).

In DNA barcoding, NGS can generate both full and partial genome sequences, enabling the creation of robust barcode markers or wider genomic amplicons for higher resolution. Among its most important applications is metagenomics, in which viral sequences are recovered from complex mixtures of environmental or host-adapted samples without culturing. This methodology is essential for virus discovery, characterization of viral diversity, and surveillance of emerging pathogens within ecosystems (Datta et al., 2015).

Additionally, NGS is highly useful in phylogenomic analysis, providing intricate details about the evolutionary relationships between viruses and their hosts. By analyzing multiple genomic locations or even full genomes, scientists can construct accurate phylogenetic trees and trace how viruses evolve, emerge, and spread. Modern virology and epidemiological research, as well as biodiversity assessment, are impossible without the scalability, precision, and versatility of NGS methodologies (Radford et al., 2012).

4.2.2.5 Multiplex Barcoding for Viral Detection and Surveillance

Advanced molecular techniques, such as multiplex barcoding, are molecular methods that enable multiple viral targets to be amplified simultaneously in the same reaction by applying uniquely labeled primers to each target sequence. This strategy significantly contributes to the efficiency and scalability of DNA-based diagnostics, with the potential to detect multiple viruses, or strains of a single virus, in a single assay. Incorporated in every primer set is a molecular barcode or identification tag that enables the amplified products to be distinguished later in downstream analyses, such as sequencing or fluorescence detection (Kiryanova et al., 2024).

The technique is particularly useful in high-throughput surveillance, where it is necessary to rapidly and cost-effectively screen many pathogens across large sample panels. For example, when used in plant health surveillance, clinical diagnostics, or epidemiological research, multiplex barcoding can simultaneously detect and monitor the occurrence of viral infections, even when pathogens are present at low titers or during co-infections. Together with platforms such as real-time PCR or NGS, multiplex barcoding can be used to diagnose diseases on a large scale, track outbreaks, and screen for biosecurity threats. It allows for optimization of reagents and processing time, as well as broad-spectrum detection, making it a crucial technique in contemporary virology and molecular diagnostics (Deshpande et al., 2016).

4.3 APPLICATIONS OF DNA BARCODING IN DNA VIRUSES

4.3.1 Viral Taxonomy and Systematics

Taxonomy and systematics of viruses are important for classifying and arranging viruses based on their genetics, structure, and functionality. As molecular methods, in particular DNA barcoding and genome sequencing, have become more advanced, the classification of viruses has shifted toward sequence-based approaches at the expense of morphology-based strategies. Semi-conserved or conserved parts of viral genomes, called genetic markers, are increasingly used to define viral species, genera, and families, especially in viruses that cannot be cultured or those with no obvious morphology (Lefkowitz et al., 2018).

Molecular methods also allow the discovery and categorization of new viruses, including those never observed in nature or in other samples and habitats, such as aquatic systems, soil, or host-associated microbiomes. By using the genetic sequences of a new isolate and comparing them with well-characterized viral sequences, scientists can accurately assign taxonomic rankings and revise virus databases maintained by authoritative organizations, such as the International Committee on Taxonomy of Viruses (ICTV). In addition to classification, viral systematics is useful for explaining evolutionary relationships between viruses. Barcoded genomic sequences, or even entire genomes, can be used with phylogenetic inference to determine the ancestry of viral lineages, their common ancestors, and their interactions with diverse host organisms or environmental conditions. This is necessary for forecasting the emergence of viruses, developing vaccines, and understanding and informing disease control efforts. Thus, viral taxonomy and systematics based on molecular data have become essential tools not only for basic research in virology but also for applied virological diagnosis (Simmonds et al., 2017).

4.3.2 Diagnostics and Disease Surveillance Using Molecular Barcoding

Molecular barcoding technology has therefore become a crucial foundation in diagnostics and disease surveillance, providing a rapid, sensitive, and specific approach with the ability to identify both emerging and known viral pathogens. DNA barcoding enables the direct identification and differentiation of viruses in clinical, agricultural, or environmental samples with high specificity and speed by targeting specific genetic markers within their genomes. This capability is particularly important in environments where prompt testing can significantly impact treatment or mitigation efforts (Choudhary et al., 2021).

DNA barcoding is also widely used in agricultural research to identify plant viruses, such as Begomoviruses, which are notorious for causing severe yield reductions in valuable crops like tomato, cotton, and chili. Early and accurate diagnosis of these pathogens enables timely management interventions, thereby minimizing economic losses and preventing outbreaks. Similarly, in veterinary and human medicine, barcoding technologies are utilized to diagnose viral infections, including influenza, coronaviruses, and zoonotic viruses, often within hours of receiving a sample (Prassan et al., 2021).

Notably, barcoding also enables surveillance schemes that allow tracking the prevalence and emergence of viruses over time and across space. This is crucial for early warning systems capable of notifying public health or agricultural authorities of an outbreak. Barcoding tools facilitate the identification of signature characteristics of viruses, including strains, mutations, and transmission methods, thereby supporting the design of informed responses, such as quarantine measures, vaccination campaigns, and vector control strategies. In general, DNA barcoding is a vital tool for protecting the health of plants, animals, and humans by enabling early and accurate monitoring of viral diseases (Tang et al., 1997).

4.3.3 Epidemiology and Phylogeography of Viruses

With the aid of molecular barcoding, enhanced sequencing, and analytical tools, molecular barcoding helps in the epidemiology and phylogeography of viruses. Using these methods, researchers can monitor the movement and changes of viral strains across both geographical locations and time. By sequencing GS virus samples collected from different locations, scientists can identify mitigation lines, virus sources, and areas where new viral strains spread into populations (Castillo et al., 2011).

Barcoded viral sequences provide epidemiological studies with genetic fingerprints to differentiate locally occurring outbreaks from imported strains. Such data are essential in structuring control measures, predicting subsequent spread, and controlling re-emergence. For example, phylogeographic studies have been employed during viral epidemics in both plant and human systems to locate sources and to better understand how environmental conditions, human behavior, or commerce influence the spread of pathogens (Trovão et al., 2015).

Furthermore, phylogeography, a form of integrative biology in which genetic data is coupled with geographic information, allows for the visualization of viral development and spread in response to ecological and demographic factors. This is particularly significant for understanding virus adaptation to new hosts, crossing species barriers, and responding to interventions such as vaccination. Overall, the application of molecular barcoding-based epidemiology and phylogeography demonstrates the potential to provide strong evidence on viral transmission processes, serving as a primary tool for predicting outbreaks, enhancing surveillance, and informing long-term disease management approaches (Wang et al., 2023).

Additionally, it combines genetic information with spatial data, enables the visualization of how viruses evolve and spread due to ecological and demographic factors. This is particularly significant in determining how viruses evolve to new hosts, acquire interspecies capability, or even react to vaccination. Overall, epidemiology and phylogeography, facilitated by molecular barcoding, promote significant insights into the dynamics of viral transmission, yielding important outcomes for prediction, surveillance, and long-term control of outbreaks (Wang et al., 2023).

4.3.4 Biodiversity and Environmental Virology

Molecular barcoding has provided a breakthrough in the investigation of viral biodiversity and environmental virology, enabling the direct detection and identification of uncultured or novel viruses in the environment, including soil, water, air, and plant surfaces, without the need for laboratory isolation. Many viruses in these ecosystems cannot be isolated using conventional methods, and molecular tools are therefore required to study, discover, and characterize them. Enriching the virome (the full viral community) of an environment can be achieved by targeting conserved genomic regions using barcoded primers or by employing the untargeted approach of metagenomic sequencing (Hejran et al., 2024).

These methods allow the resolution of viral diversity, abundance, and distribution at high resolution, facilitating the detection of previously unknown viruses and providing insight into their ecological roles. For example, viruses in soil and water play crucial roles in nutrient cycling within ecosystems, population control among microbial species, and the maintenance of healthy ecosystems. In plants, environmental virology is used to survey latent or asymptomatic viral infections, which may develop into pathogenic infections as environmental conditions change (Roossinck, 2012). Moreover, barcoded amplicons can be efficiently applied in screening large numbers of samples in biodiversity studies, while metagenomic techniques can generate complete or partial sequence data to assemble full or partial viral genomes. Applying these approaches to study viromes in extreme or poorly researched environments is particularly valuable for global viral taxonomy and the evolutionary biology of viruses. Overall, molecular barcoding in environmental virology has opened new horizons in viral ecology, facilitating both fundamental research and practical applications related to disease surveillance and ecosystem monitoring (Shestopalov et al., 2020).

4.3.5 Vaccine Development and Therapeutics

Barcoding is a molecular method vital to modern vaccine design and therapeutic strategies, as it can be used to fine-tune viral genotypes and track genetic diversity within virus populations. By performing deep sequencing of specific genome regions or whole viral genomes, investigators can identify genetic polymorphisms that affect viral pathogenicity, host interactions, and immune escape mechanisms. This comprehensive genetic information is critical for developing effective vaccines that target the most conserved and immunologically relevant regions of a virus.

Barcoding also aids in identifying viral variants that the immune system cannot detect due to mutations in antigenic sites, a common occurrence in rapidly evolving viruses, such as influenza or SARS-CoV-2 (Wang et al., 2022). Early identification of these escape variants enables scientists and public health authorities to adjust vaccine formulations and other control measures in a timely manner. Furthermore, barcoding is used to optimize therapeutic agents, including antiviral drugs and monoclonal antibodies, by defining viral genotypes that confer drug resistance (Burton and Eyre, 2021).

During preclinical and clinical studies, barcoding can be employed to assess the responsiveness of viral populations to vaccine-induced immunity, track and discriminate vaccine escape mutants, and evaluate the genetic stability of attenuated vaccine strains. In conclusion, molecular barcoding is a fundamental tool for enhancing the efficacy, safety, and adaptability of vaccines and therapeutics in the face of future viral evolution and the emergence of previously unknown diseases.

4.4 MOLECULAR BARCODING OF CIRCULAR SSDNA PLANT VIRUSES THROUGH ROLLING CIRCLE AMPLIFICATION

4.4.1 Sampling Is Simple

A very important requirement before the precise identification of viruses is the collection of plant samples and the preservation of plant DNA, particularly in tropical and subtropical countries, where most geminiviruses and nanoviruses are considered pests. The simplest method is to harvest fresh young leaves with no evidence of necrosis and place them directly on two sheets of blotting paper to dry gradually at room temperature. To prevent DNA hydrolysis-associated oxidation and deamination, the collected specimens are stored dry and in the dark at room temperature. This allows RCA/RFLP analysis to be performed on the samples years later without compromising resolution (Fischer et al., 2015; Shepherd et al., 2008). Woody perennials are among the many plant species for which this simple method has shown efficacy. The dried plant sap samples on basic paper are generally harmless because insects (aphids, leafhoppers, and whiteflies), which transmit viruses, will not feed on this material under normal circumstances. Nevertheless, more sophisticated and expensive methods of storage and sampling have been deployed, utilizing Whatman FTA™ cards, which reduce the possibility of spreading different viruses (Owor et al., 2007).

4.4.2 Extraction of Nucleic Acids

A small disk clipped from a dried leaf, for example using the lid of an Eppendorf tube, provides enough material for many RCA/RFLP analyses. For most plants, including those that are technically resistant due to mucilage and other secondary plant components, nucleic acid extraction using the cationic detergent cetyltrimethylammonium bromide (CTAB) is a reliable method (Haible et al., 2006) Plant species that exhibit browning and necrosis, including some potato and vetch varieties, may require the addition of antioxidants. The material is prepared for RCA following alcohol precipitation and chloroform extraction. Numerous attempts to improve detection by further purifying the nucleic acids to enhance reaction efficiency were unsuccessful (Haible et al., 2006). Brief heating (10 min at 65 °C) facilitates the dissolution of nucleic acids, removes any undenatured proteins, and reduces the likelihood of viral ssDNA rings breaking. A temperature of 95 °C is typically recommended in protocols to heat double-stranded DNA (dsDNA), such as plasmids, for two purposes: nicking one of the strands of DNA and denaturing it so that primers can hybridize with the remaining strand. This procedure does not work for circular viral ssDNA. Furthermore, it generates a large number of RCA templates from the host linear DNA, which must be discarded later.

4.4.3 Replication Initiated by Run-Off and De Novo Priming

Certain fundamental characteristics of geminiviruses and nanoviruses must be considered for optimal RCA performance. Circular ssDNA often constitutes the majority of viral DNA in a leaf sample. Significant amounts, however, are involved in replication, resulting in a variety of dsDNA and ssDNA forms, as well as mixtures of these two forms. Although it produces a smear signal immediately following hybridization, this DNA component is undervalued on one-dimensional gels evaluated by Southern analysis with virus-specific probes. Two-dimensional gels better illustrate the range of DNA intermediates for various geminivirus-host combinations (Jeske et al., 2001; Preiss et al., 2003; Alberter et al., 2005; Morilla et al., 2006; Jovel et al., 2007; Erdmann et al., 2010). Run-off RCA can be directly modeled by these intermediates for recombination-dependent replication (RDR), rolling circle replication (RCR), and complementary strand replication (CSR). If the specimen has been preheated to 95 °C, its direct usage is prohibited. Only the normally smaller segment of covalently closed circular (ccc) dsDNA requires nicking to begin replication. For further replication, viral ssDNA requires a primer, which may be a random hexamer added to the reaction mixture. Because of Phi29 polymerase's innate 3′–5′ exonuclease activity, the primer's 3′ end should be protected by phosphorothioate-linked nucleotides for optimal results (Dean et al., 2001).

Endogenous RNAs, such as viral transcripts and small interfering RNA (siRNA), can presumably function as primers and hybridize with ssDNA at around 65 °C, which may explain why further purification of crude nucleic acids is not always beneficial. RCA primarily yields a complex mixture of ssDNA and dsDNA, and due to Phi29 polymerase's strong processivity and strand-displacement activity, it forms a massive network. Pinwheel-like structures can be produced by priming a single circular molecule multiple times. The major ssDNA products require second-strand replication for restriction enzymes to cleave them effectively during RFLP analysis. Therefore, assay outcomes critically depend on the relative amounts of modified phosphorothioate primers and DNA template, which are typically added in excess. Infected plant samples with particularly high ssDNA concentrations may require two stages of serial dilution to achieve optimal dsDNA RCA product concentrations and distinct RFLP patterns. Partial second-strand replication is often the source of incomplete digestion, producing complex fragment patterns that can be misleading.

4.4.4 The Best Choice of Restriction Enzymes

The appropriate restriction enzymes, primarily HpaII and Sau3A, are employed depending on the virus to be identified. Enzymes that detect but do not produce sufficient fragments to distinguish among various viruses and genomic components for genomes between 2500 and 5000 nt in size are less useful (Wyant et al., 2012b). Following electrophoresis, the resultant fragments are typically separated in polyacrylamide or agarose (>5%, >2%; for <1000, >250 bps, respectively) gels and stained with SybrGold or ethidium bromide (Fischer et al., 2015; Haible et al., 2006; Jeske et al., 2010). Slower migration during longer electrophoresis times increases resolution for a wider variety of fragment lengths. The migration distances in the gel provide precise estimates of fragment lengths under ideal conditions; unusual migration behavior has been observed in only a few instances, likely due to the bent conformation of specific DNA fragments (Gutierrez et al., 1995). In such cases, the issue can be resolved by using a different restriction enzyme. Component-specific probes combined with Southern blot hybridization enhance resolution by visualizing variations in incompletely cut RCA products and viral DNA pools (Fischer et al., 2015; Jeske et al., 2014). Regardless of size, the most effective diagnostic fragments are those that vary sufficiently across separate DNA components; selecting the appropriate gel is critical. Most fragments do not meet this requirement immediately due to statistical reasons. Frequently occurring fragment lengths are more similar among viruses. Analyzing the variability in the geminivirus database with respect to fragment size and frequency is particularly informative. All GenBank geminivirus records obtained in 2017 were processed, and 9140 unique entries for different components were selected for in silico restriction analysis using Python scripts. For the most common genera (Begomovirus, Curtovirus, and Mastrevirus), nucleotide content was examined. Begomoviruses (30.1%) and curtoviruses (31.4%) exhibit a bias toward higher T frequency than expected, possibly due to excessive deamination activity causing C→T transitions, as demonstrated experimentally (Richter et al., 2016a). Restriction enzymes with A&T-rich recognition sequences are expected to produce larger and smaller fragments. In begomoviruses, these sequences are overrepresented, whereas in mastreviruses, they are underrepresented. There are only a few hotspots in begomovirus DNA regulatory elements, such as promoters (between 2500 and 200 nt) and terminators (between 1000 and 1200 nt), with distribution across the genome being mostly random. Consequently, begomoviruses can generate more and smaller fragments than mastreviruses. Comparing RFLP data across all components increases the likelihood of identifying diagnostic fragments.

4.4.5 Standardization for Determining Fragment Sizes

To distinguish between non-matching and matching fragments, as well as unique viral components, the fragment lengths obtained during gel electrophoresis must be accurately determined. The sigmoid curve provides a better match for the entire gel. However, a log connection between migration distance and molecular weight (estimated as retardation factor (Rf) scores) can be observed in the gel's center (Fischer et al., 2015; Wyant et al., 2012a). The Excel function of Excel [=5 + NORM.INV(D5;0;1)] facilitates performing a Probit analysis (Finney, 1952), which produces the inverse function for linearizing a sigmoid curve. This approach generated an error of less than 3% for fragment size measurements across a wide range of sample evaluations in the laboratory, regardless of the presence of plant viruses or gel type. It also maximized the log-lin fit ($R^2 = 0.991$) (Wyant et al., 2012a). To achieve this level of accuracy, each gel must have the proper standard fragments. Other marker restriction fragment mixes are inferior to the RFLP/RCA products of sequenced viruses when used for comparison. If a specific virus has been identified in a plant, the RCA products generated by the virus from cloned material may be adequate and straightforward to compare, allowing for calibration of the assay. When comparing unidentified viruses to an electronic database, a wide variety of fragment sizes is desirable for the standards. Because Abutilon mosaic virus (AbMV) is found in botanical and market gardens worldwide, its DNA may be selected, or the RCA products of various plasmids may be beneficial (Fischer et al., 2015). Additionally, AbMV is a valuable training tool because woody Abutilon plants contain secondary metabolites and mucilage, which can be challenging for novices to handle. The simplest technique for evaluating improper band assignment to fragment size in the gel is graphical examination. The accuracy of determining the remaining doubtful bands increases with the number of bands accurately assigned. If restriction enzyme digestion is complete, the band intensities of the relevant genomic fragments should correspond to a constant molar series for each component. Using ImageJ software to scan a lane and evaluating the decrease in peak heights allows visual verification. In addition to D-DNA products, which may be identified when samples from various plants are examined because they are often distinct in each plant, double and triple bands can be observed for fragments with comparable molecular weights (Bach and Jeske, 2014). Several known limitations regarding how individual components may be represented in RCA pools for nanoviruses have been reported (Gallet et al., 2017).

4.4.6 Get the Unknown and Unexpected

In contrast to PCR and immunodetection, RCA's primary advantage is its ability to identify circular DNA molecules without prior knowledge. The tomato yellow leaf curl virus (TYLCV) in the Americas (Duffy and Holmes, 2007) or Northern Europe (Just et al., 2014) and Squash leaf curl virus (SLCV) in the Middle East (Ali-Shtayeh et al., 2014) are examples of viruses that can be identified when introduced into another country. Previously thought to be exclusive to the Old World, samples with aSats were discovered in Venezuela, Cuba, and Brazil (Paprotka et al., 2010; Jeske et al., 2014; Romay et al., 2010). Furthermore, the mitochondrial plasmids of certain plant species were amplified (Homs et al., 2008; Paprotka et al., 2010). Important precautions are taken to prevent the transmission of geminivirus-infected material by screening germplasm bank resources using RCA/RFLP, particularly for vegetatively propagated plants (Paprotka et al., 2010).

4.4.7 Resurrection of Viruses

Simple RCA products are generally considered safe regarding natural transmission; however, infectious viruses can be created by biolistically inoculating test plants with RCA-amplified DNA (Richter et al., 2016a, 2016b; Jeske et al., 2010). This strategy also retains the entire population structure of a viral quasispecies (Jeske et al., 2010; Richter et al., 2016a). In combination with fragment selection via gel purification, circularization of DNA fragments by religation, and a second RCA, viruses otherwise resistant to bacterial cloning can achieve sufficient infectivity (Jeske et al., 2010). In rare cases, such as ACMV, rub-inoculation can render a linear genetic segment sufficiently infectious (Hipp et al., 2016). To maintain redundant sequence sections, the recombinational or replicational release of unit-length DNA circles inside cells often requires partial dimers ('bitmers'), dimers, or multimers of RCA products without additional digestion (Stenger et al., 1991). When infectious DNA is cloned using bacterial plasmids, restriction enzymes break down RCA products at specific concentrations or incubation times (Wyant et al., 2011). After gel purification, fragments larger than the genomic length are added to plasmids. Without searching for the appropriate restriction enzymes, the RCA product can typically be partially fragmented with MluI (or Sau3A) and inserted into the plasmids' EcoRI (or BamHI) sites for all genomic components. The versatile pGreen small agrobacterial plasmid (Hellens et al., 2000) has been used to transfer various viruses, including Sats, between plants via stem inoculation or leaf infiltration (Wyant et al., 2011, 2012b), while other Ti-associated plasmids are also frequently employed.

4.4.8 Get the Whole Sequence Information

In addition to traditional cloning and sequencing, direct RCA product sequencing is the preferred approach for obtaining complete sequences of viral components via primer walking (Wyant et al., 2012a). If certain sequence information allows deduction of a specific primer, and the plant being investigated is infected with a single virus, the final sequence represents the consensus or master of the DNA population, ignoring heterogeneity within the population and potentially differing from individual clones. RCA products have also been directly subjected to various deep sequencing techniques (Circomics). Longer reads are a benefit of pyrosequencing, though additional errors were noted, especially in homopolymeric stretches (Wyant et al., 2012b). When adequate plasmid standards are used to assess error rates, Illumina sequencing provides shorter but sufficient reads to identify mutational diversity within a viral DNA pool (Richter et al., 2016a). Surprisingly high percentages of D-DNAs were detected. Many distinct samples have recently been subjected to single-molecule sequencing in a single test (Mehta et al., 2017). This approach is highly useful for detecting intra- and intermolecular components of recombinants. Direct RCA product sequencing of short fragments in the most variable intergenic region of a virus component is sufficient for most applications and for identifying known viruses. For conserved primer sequences, such as those in the Rep gene or coat protein, this method outperforms RFLP analysis due to declining commercial sequencing costs. Resulting data are easily comparable to electronic databases and more accurate. Circomics expands the perspective of basic research to include viral DNA variation intricacies and the diverse chemicals produced by many plant species and varieties. Regardless of whether traditional approaches or genetic engineering are used, a deeper understanding of viral population composition has numerous implications for resistance breeding. Viruses can be considered highly polyploid organisms. Wild-type (wt) variants can complement mutants within a cell, making an error-prone lifestyle not only possible but advantageous, creating a pool of potential survivors before environmental stress occurs. Based on this principle, begomoviruses (Cleome leaf crumple virus, ClLCrV; Euphorbia yellow mosaic virus, EuYMV) and their aSats were studied in various T-DNA insertion lines of *Arabidopsis thaliana* to visualize alterations in population structure and the role of individual host genes in viral replication (Richter et al., 2016a, 2016b). Biolistic inoculation of RCA products outperformed agroinfectious clones in effectiveness and prevented bacterial mutations. Illumina sequencing identified shifting viral DNA populations, and plasmid controls verified technical error rates. High replacement rates were consistently observed.

Genetic plant lines showed minor variation, with RAD51D increasing geminiviral infection levels and KU80 slowing viral multiplication (Richter et al., 2016b). The role of error-prone translesion DNA synthesis in viral replication in phloem cells was examined (Richter et al., 2016b). Data indicated high substitution rates (10^{-4} to 10^{-3}) and biases toward nucleobase oxidations and deamination mutations. High concentrations of D-DNAs were also detected, which are generally overlooked because they represent a heterogeneous population invisible in multiple bands and only revealed by background smears in RFLP mapping. The augmentation of DI-DNAs as a cross-protective measure may be highly significant; however, as demonstrated for beet curly top virus (BCTV), D-DNAs can exhibit antagonistic effects on symptoms and infection spread (Horn et al., 2011). Consequently, using the entire viral DNA pool rather than a single infectious clone with limited field applicability is preferable when experimentally challenging a resistance trait. RCA technology also enables sustainable plant protection by monitoring changes in population structure resulting from new cultivar deployment and varying climatic conditions.

4.5 SOLID-STATE NANOPORE-BASED DNA BARCODING AND SEQUENCING

Nanopores are a well-known single-molecule analytical technique that shows great promise for genomic applications. Single-molecule techniques for biomedical applications, including genome mapping, genotyping, and DNA sequencing, have developed at an accelerated rate in recent years (Shendure et al., 2004; Harris et al., 2008; Eid et al., 2009; Milos, 2009; Singer et al., 2012; McNally et al., 2010; Rosenstein et al., 2012; Branton et al., 2008). One novel single-molecule sensing technique that hasrecently gained popularity is nanopores, which have enormous potential for high-throughput genome analysis. A key distinction of nanopores is that they do not require the probed molecules to be immobilized, as they exploit the natural charge of biomolecules, such as proteins and nucleic acids, to transport them through an applied electric field. The detection of individual DNA molecules can be achieved using this electrophoretic technique, in which DNA molecules are passed through artificial or biological nanopores (Kasianowicz et al., 1996; McNally et al., 2010).

4.5.1 Nanopore-Based Sequencing

Recognizing that one of the most ambitious NGS platforms is nanopore-based DNA sequencing (Branton et al., 2008), we introduce the latest findings below. This nanopore technology is highly useful in single-molecule genomic analysis because of two features. First, the electrophoretic focusing and threading procedures in nanopore technology are important for examining small amounts of DNA samples. Long DNA molecules are essential to be electrophoretically focused and threaded out of the bulk solution into the pore, enabling the use of small quantities of DNA. Second, since long DNA coils are frequently linearized by sub-5-nanometer pores, it is theoretically possible to employ nanopores to effectively 'scan' information along a genome (Singer et al., 2010, 2012; Wanunu et al., 2009). These properties, combined with the ability to fabricate stable solid-state nanopores in ultra-dense patterns rapidly, constitute the foundation for low-cost, high-throughput DNA sequencing (Shendure et al., 2004; Harris et al., 2008; Eid et al., 2009; Healy, 2007; Fuller e al., 2009).

Although there have been recent developments in nanopore research, two significant barriers remain for implementing a high-throughput solid-state nanopore-based device: (i) enabling parallel reading, and (ii) discriminating between four distinct nucleobases. Considering the vast length of the human genome, a method is needed to synthesize and continuously monitor a dense set of nanopores, as only a single nanopore can interrogate one DNA molecule at a time. Although the recently developed ability to distinguish nucleobases has shown promise in protein-based nanopore sequencing (Manrao et al., 2012; Cherf et al., 2012), scaling and multiplexing this technique from a single pore to a sparse array of tens of thousands of pores remains very challenging. On the other hand, solid-state nanopores, produced using standard semiconductor manufacturing techniques, are particularly suitable for parallelization in dense arrays. Their ability to distinguish the four bases in DNA has not yet been demonstrated; nevertheless, as of this writing, optipore sequencing technology was created to address both of these primary issues. The DNA of interest is transformed into an enlarged form using a highly parallel bulk procedure that sequentially replaces each nucleobase with a specific 16-mer oligonucleotide to enhance contrast between individual nucleobases. Transformed DNA is then hybridized with four-colored complementary molecular beacons to represent the four 16-mer oligos. The beacons are sequentially removed, or unzipped, using nanopores to detect the sequence. A CCD or complementary metal-oxide semiconductor (CMOS) camera records the four-color photon bursts produced when a fluorophore is unquenched with each unzipping event. The original DNA sequence can be reconstructed from the registered photon bursts. To reduce the DNA translocation rate to a speed suitable for observation by single-molecule optical techniques, the DNA unzipping mechanism is operationalized, which is hindered by the rapid motion of DNA through nanopores (McNally et al., 2008; Soni et al., 2010). This procedure can be multiplexed using optical detection; theoretically, a single camera could probe an array with thousands of nanopores simultaneously. Mostly, nanopore-based sequencing applies the electrophoretic force (adjusted by voltage or other easily controlled physical parameters) to drive DNA strands through a pore. This concept underlies 'physical' methods for sequencing DNA through artificial nanopores, such as the OptiPore method described here. Each DNA base is biochemically transformed for direct reading via an unaltered solid-state nanopore. Thus, readout length and speed are not restricted by enzymes. The use of optical sensing instead of electrical sensing enables an easy transition to multipore detection for DNA sequence identification. Our approach uses total internal reflection to achieve high spatio-temporal resolution optical detection of individual DNA molecules translocating through a nanopore (Soni et al., 2010).

4.5.2 Cyclic DNA Conversion

Cyclic DNA conversion (CDC) is a biological method for increasing the 'footprint' of single nucleobases in a nucleic acid molecule. In CDC, pre-made oligonucleotides replace each nucleobase in a given DNA sequence. These pre-made oligonucleotides, called 'code oligos,' are usually 16-mers long and contain sequences selected to promote optimal conversion in the CDC procedure. Each of the four natural nucleobases is represented by one of the four 'code oligos' developed. One nucleobase or several nucleobases are replaced with their corresponding 'code oligo' during a single conversion cycle. Multiple conversion cycles allow concatenation of several 'code oligos' into a lengthy ssDNA molecule readable by a nanopore-based DNA sequencer. After conversion, the single DNA strand is hybridized with fluorescently tagged, complementary, predesigned oligonucleotides. These oligonucleotides form beacons for optical detection. Target DNA amplification is not required in CDC, avoiding errors from polymerase-based amplification and eliminating further purification of the converted products. This method supports complete integration into lab-on-a-chip systems because only minimal DNA material or reagents are needed. The procedure begins with preparation of target ssDNA. Genomic DNA is enzymatically fragmented into blunt-ended short fragments (around 1 kb), and one end is sealed by ligating a biotin-labeled adaptor. Once the strands are immobilized and captured on magnetic beads, DNA melting separates them. Target ssDNA is then subjected to multiple circular conversion steps. In the first step (i), a library of T-shaped oligonucleotide probes with asymmetric single-stranded overhangs on both sides of a central base-paired region is hybridized. A 5′ site (yellow, labeled n) contains all four viable bases, and the 5′ overhang (red) fully complements the corresponding tag on the target ssDNA. The red portion of the double-stranded region indicates the code oligo of the specific base. The black component of the three single-stranded overhangs identifies the nucleobase (A, G, C, or T). The library contains all four nucleotides to cover all positions of the 3′ overhang (illustrated as n). Step (ii) uses a ligase reaction to ensure high hybridization fidelity of the probe library, demonstrating superior reliability compared to other enzymatic processes (Shendure et al., 2005). Step (iii) employs a type IIS restriction enzyme to cleave off the recognition site. Step (iv) removes the probe's top section following brief temperature-induced melting. Completion of these four steps moves a single base from the target ssDNA's 5′ end to its 3′ end, adding the corresponding code along with additional bases that serve as the restriction enzyme type IIS recognition site (blue). The system is then ready for the next cycle. In theory, long reads can be converted by repeating this process as needed. Each cycle is independent of previous cycles, making CDC intentionally Markovian. Inaccurate hybridizations are eliminated in subsequent cycles and are not ligated. Template conversion is fully asynchronous. Since the transformed DNA is concatenated from 'bits' designed to minimize cross-hybridization, a stable secondary structure is not expected. With conversion time as the primary constraint, lengthy DNA strands can be converted efficiently.

4.5.3 Optical Detection

For simplicity, the optical detection scheme's proof-of-principle was first tested with two colors. In this case, bit '0' and bit '1' are two-bit combinations of two specified, distinct sequences that represent the four nucleobases. In this manner, we assigned A to '1 1,' G to '1 0,' T to '0 1,' and C to '0 0.' The '0' and '1' sequences are complementary to and specifically hybridize with two distinct molecular beacons containing two different kinds of fluorophores. The digitalized hybridization of DNA molecules results in the two distinct molecular beacons following the conversion of the target DNA. Following hybridization, the beacons are gradually removed from the dsDNA molecules as they are electrophoretically threaded through a solid-state channel about 5 nm in diameter. Every time a molecular beacon is removed, a fresh fluorophore is unquenched, producing a photon burst that is captured at the pore's location using an electron multiplying CCD (EM-CCD), as previously explained.

The two-color coding of the photon bursts uses red and blue, and the positional pattern at the pore location corresponds to the binary code of the target DNA sequence. This demonstrates the utilization of the OptiPore preparation, in which we used our own developed total internal reflection imaging technique with solid-state nanopore technology to distinguish the four DNA nucleotides. To differentiate the four nucleotides, a two-color coding system was created using two high-quantum-yield fluorophores, Atto647 (A647) and Atto680 (A680). The optical emission signal could also be directed using a dichroic mirror into a dual channel, which is subsequently recorded in parallel using an EM-CCD camera. Since both fluorophores have overlapping spectra, there is some cross-leakage between the emissions of the two dyes: the A647 emission bleeds into the A680 (channel 2) channel and vice versa. Consequently, it was determined that, instead of using the relative intensities in each channel, the ratio (R) between the fluorescence intensities of channel 2 of the fluorescent molecular beacons and channel 1 can be used as a robust method to differentiate the two fluorescent molecular beacons. For A647 and A680, calibration unzipping measurements using each fluorophore independently produced R values of 0.20 ± 0.06 and 0.40 ± 0.05, respectively. Four distinct two-beacon combinations were determined using the intensity ratio calibration distributions: 10 (G), 11 (A), 01 (T), and 00 (C). The A647 and A680 beacons are represented by the numbers 0 and 1, respectively. At 0.21 ± 0.05 and 0.41 ± 0.06, the bimodal distribution of R shows two modes that are fully consistent with the calibration values. To identify the base, all photon bursts with $R < 0.30$ were classified as '0,' whereas those with $R > 0.30$ were classified as '1.' Additionally, the probability of proper classification was calculated using the R distribution. As a result, the two channels can be statistically calibrated to achieve the best possible discrimination between the two fluorophores. All four DNA bases can be identified as singlets, as shown by an example of an optical unzipping event ($n > 2000$ photon bursts). The high ratio of intensity (R) between the two channels and the high signal-to-background ratio of photon bursts are major contributors to the effectiveness of two-color recognition. Using the calibration distributions, a novel technique was developed to automatically detect photon bursts and identify beacons. Base calling was performed using this algorithm with matching certainty scores. The categorization error per nucleobase is below 10% (per single read) even in the direct experimental configuration used here, where only one excitation line and commercially available agents were applied. These findings demonstrate that solid-state nanopores can recognize the nucleobases of transformed DNA. For a single read, the raw error in base calling is relatively small, and this error is expected to decrease significantly with further refinement of the technique. Specifically, the addition of four distinct color beacons is expected to double the sequencing speed by drastically reducing base calling uncertainty to below 1%. Optically based nanopore DNA sequencing offers several advantages over other techniques. The categorization error per nucleobase is less than 10% (per single read), even in a direct experimental setup with only one excitation line and a single commercially available agent. Furthermore, the readout stage of the nanopore device does not involve enzymes; instead, it is a straightforward array of solid-state pores. The readout speed is fully adjustable and limited solely by the temporal resolution of the sensing modality. Compared to other single-molecule methods, the current reading speed of 50–250 bases per nanopore per second is favorable (Manrao et al., 2012; Cherf et al., 2012). Using improved reagents and a simple design enhancement to four-color detection, this speed is expected to exceed 500 bases per second per nanopore, resulting in a raw base calling error of less than 1%. Importantly, optical readout enables the use of commercially available high-sensitivity cameras to read directly from hundreds of pores.

4.5.4 Genotyping Using Nanopore Technology

The capacity of solid-state nanopores to quickly and reliably probe individual RNA or DNA molecules isolated from a very tiny sample pool, down to a few attomoles (about 10^5 copies), sets them apart from previous single-molecule approaches (Wanunu et al., 2009). These special characteristics make solid-state nanopores one of the best candidates to be used in molecular diagnostics, especially where the number of cells used is small. Despite these encouraging characteristics, nanopores have not yet been widely used in molecular diagnostic sensing applications, mainly because of their inability to selectively detect specific DNA or RNA sequences. However, through the use of tagged-DNA probes (Rapireddy et al., 2007; Nielsen, 2010; Sahu et al., 2011; Chenna et al., 2008) and nanopores, it has just been possible to positively identify and differentiate between two genes with very close sequence similarity. This can now be demonstrated as being able to identify and differentiate two forms of the same gene between two very close subtypes of the human immunodeficiency virus (HIV), thus providing a new platform for the classification of viruses. Specifically, we showed that probe sites on sub-millisecond timescales could be resolved with great ease down to short distances (of the order of a few tens of nanometers). The incorporation of PNA probe libraries complements the inherent benefits of solid-state nanopore technology, enabling an electronic diagnostic platform based on nucleic acids that was previously unheard of in terms of speed. PNA probes are new, synthetic oligomers that have exceptionally high affinity for DNA; they insert themselves into the DNA-DNA helix with relatively high specificity and are restricted by sequence, replacing one DNA strand with which they form a complex. Surprisingly, every one of the labeled parts of larger cross-section directs very specific and individual ion current blockades when directing PNA-tagged DNA strands through a nanopore. One example of such a discovery is the pre-selected DNA molecule consisting of three PNA binding sites. In the absence of PNA probes, the translocations of DNA molecules demonstrate discrete blockades of the current with characteristic dwell times and amplitudes, and a single value of blocked current that is very close to the average cross-section of the dsDNA (Wanunu et al., 2008). With each event of translocation, three additional instances of blockages are then observed after invasion of the PNA probes, and all these occur according to a particular and constant ion-current condition. The advances made in our study form the foundation of a new single-molecule gene identification method, given that we can now target and characterize short sequences (i.e., 15 bp) within any DNA molecule of interest. The last step beyond counting the number of PNA/DNA sites per target molecule will necessitate the facility to read out the specific sites of the PNA/DNA sites on the DNA molecule to barcode the target without an ambiguous exception. That is, the spacing of the locations of probes, as well as the number, must match a specific pattern of a given gene in order to be sure of identifying that gene. To evaluate the accuracy of nanopores in resolving the separation of two PNA probes along a DNA molecule, six randomly oriented DNA molecules with two copies of identical PNA sites were produced. The length and sequence of the two flanking regions (approximately 1200 bp) were identical in all samples, and the distance n between the two PNA binding regions ranged between 100 and 1000 bp. The mean value of t, measured using N = 2560 events per molecule of the six molecules, was plotted as a function of the actual gap distance between the two PNA probe sites (n) in nanopore experiments using these molecules, in order to determine the delay time between pairs of PNA probes. It was able to specifically recognize the shortest n (100 bp), which is likely to form a distance of 34 nm. In addition, we found that a power-law relation with exponent 1.39 +/– 0.09 best described our results, which was also in agreement with other reports that had previously measured the translocation time of untagged DNA molecules of similar length (Wanunu et al., 2008). This calibration process enables a PNA binding field down to a few probe-to-probe spacings on a DNA strand, including a nanopore readout signal far enough to read it. Theoretically, a reasonably modest probe library suitable for a wide range of genomic targets can be created by combining data on the distance among probes bound and the number of probes. The positive attribute of the relative ease of the use of high sequence specificity of 15-mer PNA probes, and the potential to accurately map relative probe positions along DNA using the nanopore method in a simple barcoding strategy, is leveraged in assessing the feasibility of the method as a potential molecular diagnostic tool. To carry out this process, the DNA sample is extracted from the source of interest and placed in a situation where it can mix with a set of probe libraries. What occurs is that the probe sequences that do not possess a complementary sequence to the target sequences will not bind to the extracted DNA. This concept allows us to create so-called barcode maps of possible binding patterns that are specific to a given library of different PNA probes. A pattern and delay durations of PNA were measured in a nanopore. Electronic barcoding of single DNA molecules using a PNA technique (or peptide nucleic acid) is shown. A probe DNA example of a 3-PNA (three-probes nanopore DNA) is aligned symmetrically on the two sides of 4-nm solid-state nanopores drilled on a thin layer of silicon nitride membrane. As the altered synthetic PNA-tagged region passed through the nanopore, it was characterized by three distinct secondary blockade events, which were beyond the usual reading of DNA current blocking. The inset on the top represents a transmission electron microscope (TEM) picture of the thin layer of silicon nitride through which the 3.7-nm nanopore was created. Bottom inset: an approximate close-up structure at a PNA probe site.

The single-stranded PNA (red) has a strong affinity for DNA and binds to its corresponding DNA sequence, displacing the second DNA strand locally. The helical structure undergoes a local alteration as a result of this action, making it distinguishable by the solid-state nanopore. Publications from CACS. The gene that codes for reverse transcriptase, integrase, and protease is a prime target, as it is essential to the creation of an HIV vaccine (Gulick, 2010). Additionally, this gene was a strong contender for a proof-of-concept study due to the high pol similarity across the subtypes (92% identity after aligning the chosen subtypes). A subsequent comparison can then be made with these maps regarding the presence or absence of the desired DNA molecules. As an exemplary sample set used to analyze this approach, HIV genomes were used. HIV can serve as a model organism to test this methodology, since the virus naturally has the ability to rapidly become resistant to treatment. Consequently, a simple, cheap, and fast method to detect HIV drug-resistant issues has to be devised (Gulick, 2010; Mabey et al., 2004; Thompson et al., 2010). The HIV pol gene, which encodes proteins including protease, integrase, and reverse transcriptase, is an ideal target since it is essential in the development of an HIV vaccine (Gulick, 2010). In addition, the gene was also a challenging candidate for a proof-of-concept study, considering the high levels of pol conservation in the subtypes (92% identity after alignment of the chosen subtypes).

The DNA fragments along with the pol gene sequences, represented by 3050 bp of HIV-1/C and HIV-1/B, were constructed based on standard procedures, and they were at least 90% similar (Rodenburg et al., 2001). Four PNA oligomers were also produced, two of which (PNA1 and PNA2) would be exactly complementary to both subtypes, the third one (PNA3) would match only the sequence found in subtype B, and the fourth one (PNA4) would match only the sequence present in subtype C. This should produce two different patterns or barcodes: PNA1 and PNA2 to be at least 850 bp apart in both subtypes, PNA3 to be at least 850 bp apart in subtype B, and PNA4 450 bp apart in subtype C. After selective cleavage of each gene into 10 fragments of 300 to 400 bp each and gel-shift experiments with each of the four specific PNA oligos, both separately and in combination as a complete set, specific PNA binding profiles were confirmed (Singer et al., 2012). Such a labor-intensive gel-based study can be superseded by a single nanopore reading of every one of the two types of HIV genes and can take only a few minutes.

First, we noticed that the minor sequence variations of the two subtypes are not sensitive to nanopore analysis of the untagged subtypes. A simple comparison of the recorded delay durations between the three independent PNA sites allowed a clear observation that both subtypes, as PNA-tagged DNA, produced distinctive ion current signals. Standardization of the individual delay times of PNA sites with reference to aggregate delay times of all the delay events in each of the events made it simple to investigate thousands of molecules. This process eliminated small pore-to-pore fluctuations, making the data much clearer and easier to interpret. We calculated a delay time t12 between the first and the second PNA sites, and t23 between the second and the third PNA sites. In subtype B and subtype C (red and blue, respectively, with N = 750 events), one defines normalized delay times as 1 = t12 / (t12 + t23) and 2 = t23 / (t12 + t23). Because the delay phases between the PNA probes are roughly equal (850 bp), it is not surprising that the symmetrically labeled subtype B mutant yields two distributions that are nearly the same, with an intermediary peak at 0.500 rather than the central point at 0.500. On the other hand, the subtype C variation gave double peak distributions with centers occurring at 0.252 ± 0.004 and 0.731 ± 0.005, and 0.269 ± 0.005 and 0.749 ± 0.004 in 2 and 1, respectively.

Because 0.61 / (0.61 + 1.85) = 0.248 and 1.85 / (0.61 + 1.85) = 0.752, these values exactly match the delay period linked to the 450- and 850-bp gaps, as seen in the calibration curve: 0.61 ms and 1.85 ms. The two nearly similar distributions in each instance reflect the DNA's ability to pass through the nanopore with either end entering first. More significantly, this implies that the normalized delay periods between PNA signals are a reliable way to measure the linear distance between target sequences along the DNA. As few as four PNA probes are required to identify HIV subtypes C and B, which exhibit sequence variation of less than 8%. However, it is theoretically possible to use nanopore-based sensing for considerably longer DNA molecules with more tags, representing far more complex genomes (Wanunu et al., 2009). Moreover, we can significantly lower the possibility of inappropriate categorization by including more PNA sites and using both the spatial information (as done here) and the number of sites. Therefore, by establishing a spatially resolvable pattern of γPNA sites, the technique we present can be utilized to identify and distinguish across genomes with comparatively little sequence variation. Recent technological developments have made it possible for disease recognition methods to move from antibody-associated assays to nucleic acid-dependent assays, enabling the detection of genotypes. These developments will also help these methods continue to evolve from testing large laboratory samples to point-of-care diagnostics (Liu, 2008). The sub-classification of disorders with similar pathophysiological characteristics but distinct treatment responses will be possible with further developments in molecular diagnostics (Green et al., 2011). A purely electrical readout combined with a genomic barcoding method appears feasible based on the results provided here and the scalability of this procedure. An unknown target can be swiftly probed to ascertain its precise nature using the γPNA/nanopore approach. This method uses highly specific probes, termed γPNA probes, which are more sensitive to sequence mismatches than DNA probes, and can be multiplexed with γPNA libraries to provide sequence specificity (Karkare and Bhatnagar, 2006; Zhang and Appella, 2010). Solid-state nanopores are among the most sensitive single-molecule techniques designed so far and are implemented in this technique to fulfill its high-throughput, small-sample requirements, and resolution. Combining these techniques provides the foundation for a revolutionary diagnostics platform that can make it easier to customize a plan for the treatment of patients according to the specific variant or associated subtype of their infections. These technological advances could result in significant improvements in patient quality of care, as well as a decrease in healthcare expenses.

4.6 CHALLENGES AND LIMITATIONS OF DNA BARCODING OF VIROLOGY

Despite the significant advances in DNA barcoding for virology, several challenges limit its efficiency, accuracy, and universal application. Unlike multicellular organisms, viruses show high rates of mutation, have small genome sizes, and lack specific molecular markers, which makes barcoding of viruses more difficult. While DNA barcoding could be useful for finding viruses, various difficulties need to be overcome. Such traits make it harder to accurately identify and classify organisms. Improving the use of barcodes, updating the reference databases, and adding advanced molecular methods are necessary to increase accuracy. It is very important to focus on these problems to improve the use of DNA barcoding in virology.

4.6.1 Biological Limitations: High Rates of Mutation

DNA viruses that have high mutation and recombination rates make it difficult to use the DNA barcode method for their identification and classification. As a result of these genetic variations, it can become difficult to use DNA barcoding to track different strains of the virus. Most DNA viruses are considered slow to evolve, probably because of their DNA polymerases, which can fix mistakes during genome replication, but some groups of single-stranded DNA viruses actually show high levels of mutation and recombination (Thielecke et al., 2017).

Recombination occurs at a high rate in adenoviruses, papillomaviruses, and geminiviruses, which have regions from different evolutionary lineages. Barcode-based identification proves difficult with such mosaicism because recombinant sequences may not be well aligned with standard barcodes, which may lead to incorrect classifications and errors in the study of epidemiology and viral taxonomy (Grigoras et al., 2014).

One of the major challenges regarding DNA barcoding in virology is the very fast evolution within viral populations. Although DNA viruses are considered more stable than RNA viruses, they tend to generate sufficient genetic variation to pose as new emerging variants within short evolutionary timeframes. Due to these genetic changes, the differences between viral 'species' and 'strains' are unclear, which affects how species can be defined using sequence information alone. In addition, the high frequency of recombination enables the formation of chimeric genomes that do not conform to existing phylogenetic locations, further complicating barcode-based classification (Joly et al., 2014).

When the processes of mutation and recombination take place in viruses, genetic diversity increases, forming what is called quasispecies. This phenomenon poses significant complications for the standardization of barcodes. For example, viruses such as HIV-1 and hepatitis C virus (HCV) consist of different genetic forms within the bodies of each host. Furthermore, SARS-CoV-2 develops around 1–2 mutations every month, so the virus is always producing new variants that hinder targeting them with strain-specific data. Retroviruses, which often change their genetic structure, lead to the formation of chimeric genomes that are difficult to classify using conventional techniques. As a result of high mutation and recombination rates, DNA barcoding is less reliable and accurate for the identification, classification, and surveillance of viruses in clinical and environmental settings (Zuo et al., 2024).

4.6.2 Primer Design and Amplification Challenges

The first step of DNA barcoding is to design a primer and then amplify it successfully. This step is important because viral genomes are known for high mutation rates, frequent recombination, and a more compact structure—these characteristics tend to cause major difficulties for DNA barcoding of viruses. As a result, it becomes difficult to design universal or species-specific primers that reliably amplify the selected barcode region among different groups of viruses (Vernet, 2004).

It is very challenging to design primers for DNA barcoding of viruses because different viral families often lack conserved sequences. The situation is different for bacteria and fungi, since their ribosomal RNA genes usually have conserved sequences where primers can bind, while there are fewer such sequences in viruses. Due to this, it is difficult to design primers that work well across multiple taxa. In fact, even between viral species within the same viral family, there can be significant sequence variability, leading to poor primer binding, low amplification efficiency, or PCR failure (Houldcroft et al., 2017).

Furthermore, viral genomes are often short and densely packed with overlapping open reading frames (ORFs). This restricts the amount of unique DNA regions available for primer binding due to the genomic architecture. Overlapping primer designs may mistakenly amplify various regions or cause non-specific amplicons to form. Viral genomes are so compact that there is often little extra DNA where primers can be attached without interfering with coding or regulatory regions (Dadonaite et al., 2019).

The existence of host DNA or environmental DNA in test samples increases the complexity, making primer design more challenging. Since viruses are often found in mixed groups of microbes or within hosts, primers must be checked to ensure they do not react with the DNA of hosts or other microorganisms. Non-specific amplification can sometimes cause downstream sequencing methods to select the wrong DNA, especially if the original viral DNA constitutes only a small part of the total nucleic acid tested. Template quality and quantity may vary significantly depending on the type of sample used (clinical, environmental, plant, or insect). Many DNA viruses are present in small amounts, and their DNA can be degraded by environmental factors or enzymes within the host. These issues may decrease the effectiveness of PCR amplification, even when using well-designed primers (Martinelli et al., 2015).

4.6.3 Limitations of Reference Databases

The success of DNA barcoding strongly depends on having detailed and high-quality reference sequence databases. However, viral reference databases, such as GenBank and ViPR (Virus Pathogen Resource), have several limitations.

Incomplete reference database: Many viruses, mostly from non-model animals, environmental sources, or newly discovered viruses, have not yet been fully profiled or have no sequence information. Because reference data are incomplete, the accuracy of barcode-based identification is affected, especially when applied to novel or very different viruses. These databases are less useful for identifying organisms when inaccurate annotations, overlapping entries, or inconsistent data are present. Specifically, the Circoviridae and Nanoviridae virus families exhibit great genetic variation and often lack standardized barcode regions (Jeske, 2018).

Having similar sequences submitted multiple times can inflate the data on a few viral groups, which may mislead phylogenetic analyses. Some databases might still include low-quality or potentially incomplete sequences, including partial genomes. Some viral sequences in databases do not provide important information, such as host, geographic location, isolation procedure, or the date the sample was collected. If standardized approaches to collecting metadata are not used, it becomes more challenging to track the evolution of viruses, understand host–pathogen interactions, and explore ecological changes (Goodacre et al., 2018).

4.6.4 Bioinformatics and Data Analysis Challenges

The accurate processing and interpretation of barcode sequences in virology present significant bioinformatics challenges. An analysis pipeline usually includes quality control, sequence alignment, taxonomic classification, and phylogenetic inference. All stages are influenced by the selection of software tools, reference databases, and algorithmic parameters, whose choice may significantly affect the final output of barcoding. Major bioinformatics challenges have been encountered in the processing and interpretation of barcode sequences in the context of virology. In most cases, the analysis pipeline includes quality control, sequence alignment, taxa identification, and phylogenetic inference. The recognition obtained through barcoding relies heavily on the timely selection of software, databases, and parameters at each step (Lambert et al., 2018).

One major problem in viral barcoding is the lack of a standard method for defining species when comparing barcode samples. Although the 97%–99% similarity level is typically used for bacterial species classification in 16S rRNA genes, this rule does not apply to viruses. Because viral taxonomy is complex, it normally requires specific thresholds to be set at the family or genus level. Therefore, the same DNA sequence can be classified differently depending on the type of analysis performed (Marz et al., 2014).

Viral barcode analysis is also made more challenging by various technical issues. Clinical or environmental samples usually contain only small amounts of DNA viruses, surrounded by a variety of DNA from the host and other microbes. As a result, viral DNA is sometimes difficult to extract and amplify, causing unwanted artifacts in the sequencing process, such as chimeras, incorrect assemblies, and amplification bias (Rodríguez et al., 2009).

4.7 CONCLUSION

The application of DNA barcoding in the field of virology has enhanced capabilities for detecting, classifying taxonomy, conducting epidemiological studies, and assessing viral biodiversity. Using short and standardized genomic sequences as molecular identifiers, DNA barcoding is a sensitive, rapid, and scalable method of differentiating viral species and within-species viral strains, including in complex samples that cannot be cultured. This method has been overcoming challenges associated with the application of classical methods of identification, such as morphology, serology, or host-based methods. This chapter has been devoted to investigating the principles, procedures, and methods related to DNA barcoding, including marker choice, primer design, amplification methods, and techniques such as NGS and Nanopore sequencing. These technologies have been very useful in enhancing the efficiency and accuracy of viral detection, thus identifying both the known and novel viruses' DNA in diverse hosts and environments. Moreover, the importance of barcoding in both practical and theoretical virology can be determined by its integration in diagnosis, environmental control, and virome surveillance. Regardless of these advancements, the high mutation and recombination rates among viruses, the absence of universally conserved genomic regions across viral families, and the issue of overlapping reading frames make barcode uniformity a challenge. Furthermore, the accuracy and universality of DNA viral barcoding are impacted by limited reference databases, amplification bias, and primer design complications. It offers a single, integrated system to enhance worldwide efforts in virus taxonomy, track the pathways of viral transmission, identify infectious agents, and assess the diversity of viruses across clinical, agricultural, and ecological domains. Its contribution to the early detection of outbreaks, outbreak-responsive vaccine development, and enhanced pathogen surveillance aligns with major health priorities and initiatives, such as One Health and the Global Virome Project (GVP). In conclusion, the advantages of DNA barcoding stem from its capability to enhance viral research and public health strategies. Although some challenges remain, ongoing technological advancements and collaborative work between virologists, taxonomists, and bioinformaticians will streamline and increase the applicability of this methodology. The application of barcoding in virology enhances the efficiency of virus identification, facilitating a deeper understanding of their evolutionary, ecological, and systemic implications globally.

REFERENCES

Alberter, B., Rezaian, A.M., Jeske, H. 2005. Replicative intermediates of ToLCV and its satellite DNAs. *Virology* 331, 441–448.

Ali-Shtayeh, M.S., Jamous, R.M., Mallah, O.B., Abu-Zeitoun, S.Y. 2014. Molecular characterization of watermelon chlorotic stunt virus (WmCSV) from Palestine. *Viruses* 6, 2444–2462.

Antil, S., Abraham, J.S., Sripoorna, S., Maurya, S., Dagar, J., Makhija, S., Bhagat, P., Gupta, R., Sood, U., Lal, R. 2023. DNA barcoding, an effective tool for species identification: A review. *Molecular Biology Reports* 50, 761–775.

Aylward, F.O., Moniruzzaman, M. 2022. Viral complexity. *Biomolecules* 12, 1061.

Babaian, A., Edgar, R. 2022a. Ribovirus classification by a polymerase barcode sequence. *The Journal of Life & Environmental Sciences* 10, e14055.

Bach, J., Jeske, H. 2014. Defective DNAs of beet curly top virus from long-term survivor sugar beet plants. *Virus Res.* 183, 89–94.

Barzon, L., Lavezzo, E., Militello, V., Toppo, S., Palù, G. 2011. Applications of next-generation sequencing technologies to diagnostic virology. *International Journal of Molecular Sciences* 12(11), 7861–7884.

Becherer, L., Borst, N., Bakheit, M., Frischmann, S., Zengerle, R., von Stetten, F. 2020. Loop-mediated isothermal amplification (LAMP)–review and classification of methods for sequence-specific detection. *Analytical Methods* 12(6), 717–746.

Beerenwinkel, N., Günthard, H.F., Roth, V., Metzner, K.J. 2012. Challenges and opportunities in estimating viral genetic diversity from next-generation sequencing data. *Frontiers in Microbiology* 3, 329.

Blundell, J.R., Levy, S.F. 2014. Beyond genome sequencing: Lineage tracking with barcodes to study the dynamics of evolution, infection, and cancer. *Genomics* 104, 417–430.

Branton, D., D. W. Deamer, A. Marziali, H. Bayley, S. A., Benner, T. Butler and J.A. Schloss. 2008. The potential and challenges of nanopore sequencing. *Nature biotechnology* 26(10), 1146–1153.

Burton, T.D., Eyre, N.S. 2021. Applications of deep mutational scanning in virology. *Viruses* 13(6), 1020.

Carroll, D., Daszak, P., Wolfe, N.D., Gao, G.F., Morel, C.M., Morzaria, S., Pablos-Méndez, A., Tomori, O., Mazet, J.A. 2018. The global virome project. *Science* 359, 872–874.

Castillo, X.O., Fermin, G., Tabima, J., Rojas, Y., Tennant, P.F., Fuchs, M., Sierra, R., Bernal, A.J., Restrepo, S. 2011. Phylogeography and molecular epidemiology of Papaya ringspot virus. *Virus Research* 159(2), 132–140.

Chenna, V., S. Rapireddy, B. Sahu, C. Ausin, E. Pedroso and D. H. Ly. 2008. A simple cytosine-to-G-clamp nucleobase substitution enables chiral γ-PNAs to invade mixed-sequence double helical B-form DNA. *Chembiochem: a European journal of chemical biology* 9(15), 2388.

Cherf, G. M., K. R. Lieberman, H. Rashid, C. E. Lam, K. Karplus and M. Akeson. 2012. Automated forward and reverse ratcheting of DNA in a nanopore at 5-Å precision. *Nature biotechnology* 30(4), 344–348.

Choudhary, P., Singh, B.N., Chakdar, H., Saxena, A.K. 2021. DNA barcoding of phytopathogens for disease diagnostics and bio-surveillance. *World Journal of Microbiology and Biotechnology* 37, 1–20.

Dadonaite, B., Gilbertson, B., Knight, M.L., Trifkovic, S., Rockman, S., Laederach, A., Bauer, D.L. 2019. The structure of the influenza A virus genome. *Nature Microbiology* 4(11), 1781–1789.

Datta, S., Budhauliya, R., Das, B., Chatterjee, S., Veer, V. 2015. Next-generation sequencing in clinical virology: Discovery of new viruses. *World Journal of Virology* 4(3), 265.

Dean, F.B., Nelson, J.R., Giesler, T.L., Lasken, R.S. 2001. Rapid amplification of plasmid and phage DNA using phi 29 DNA polymerase and multiply-primed rolling circle amplification. *Genome Res.* 11, 1095–1099.

Deshpande, A., McMahon, B., Daughton, A.R., Abeyta, E.L., Hodge, D., Anderson, K., Pillai, S. 2016. Surveillance for emerging diseases with multiplexed point-of-care diagnostics. *Health Security* 14(3), 111–121.

Duffy, S., Holmes, E.C. 2007. Multiple introductions of the old world begomovirus tomato yellow leaf curl virus into the new world. *Appl. Environ. Microbiol.* 73, 7114–7117.

Eid, J., A. Fehr, J. Gray, K. Luong, J. Lyle, G. Otto and S. Turner. 2009. Real-time DNA sequencing from single polymerase molecules. *Science* 323(5910), 133–138.

Epstein, J.H., Anthony, S.J. 2017. Viral discovery as a tool for pandemic preparedness. *Rev. Sci. Tech.* 36, 499–512.

Erdmann, J.B., Shepherd, D.N., Martin, D.P., Varsani, A., Rybicki, E.P., Jeske, H. 2010. Replicative intermediates of maize streak virus found during leaf development. *J. Gen. Virol.* 91, 1077–1081.

Finney, D.J. 1952. *Probit Analysis*. Cambridge University Press: London, UK.

Fischer, A., Strohmeier, S., Krenz, B., Jeske, H. 2015. Evolutionary liberties of the Abutilon mosaic virus cluster. *Virus Genes* 50, 63–70.

Fitzmeyer, E.A. 2024. *Linking Mosquito Midgut and Virus Population Biology at the Molecular and Cellular Level*. Colorado State University.

Fitzmeyer, E.A., Gallichotte, E.N., Ebel, G.D. 2023. Scanning barcodes: A way to explore viral populations. *PLoS Pathogens* 19, e1011291.

Fuller, C. W., L. R. Middendorf, S. A. Benner, G. M. Church, T. Harris, X. Huang, and D. V. Vezenov. 2009. The challenges of sequencing by synthesis. *Nature biotechnology* 27(11), 1013–1023.

Gallet, R., F. Fabre, Y. Michalakis and S. Blanc. 2017. The number of target molecules of the amplification step limits accuracy and sensitivity in ultradeep-sequencing viral population studies. *Journal of virology* 91(16), 10–1128.

Ghosh, S., Brown, A.M., Jenkins, C., Campbell, K. 2020. Viral vector systems for gene therapy: A comprehensive literature review of progress and biosafety challenges. *Applied Biosafety* 25, 7–18.

Ginzinger, D.G. 2002. Gene quantification using real-time quantitative PCR: An emerging technology hits the mainstream. *Experimental Hematology* 30(6), 503–512.

Goodacre, N., Aljanahi, A., Nandakumar, S., Mikailov, M., Khan, A.S. 2018. A reference viral database (RVDB) to enhance bioinformatics analysis of high-throughput sequencing for novel virus detection. *MSphere* 3(2), 10–1128.

Green, E., M. Guyer and National Human Genome Research Institute. 2011. Charting a course for genomic medicine from base pairs to bedside. *Nature* 470, 204–213.

Grigoras, I., Ginzo, A.I.D.C., Martin, D.P., Varsani, A., Romero, J., Mammadov, A.C., Gronenborn, B. 2014. Genome diversity and evidence of recombination and reassortment in nanoviruses from Europe. *Journal of General Virology* 95(5), 1178–1191.

Gulick, R. M. 2010. Antiretroviral treatment 2010: progress and controversies. *JAIDS Journal of Acquired Immune Deficiency Syndromes* 55, S43–S48.

Gutierrez, C., Suarez-Lopez, P., Ramirez-Parra, E., Sanz-Burgos, A., Poenninger, J., Xie, Q. 1995. DNA bending as a potential regulatory cis-acting element of the geminivirus intergenic region. *Agronomie* 15, 415–420.

Haible, D., Kober, S., Jeske, H. 2006. Rolling circle amplification revolutionizes diagnosis and genomics of geminiviruses. *J. Virol. Methods* 135, 9–16.

Harris, T. D., P. R. Buzby, H. Babcock, E. Beer, J. Bowers, I. Braslavsky and Z. Xie. 2008. Single-molecule DNA sequencing of a viral genome. *Science* 320(5872), 106–109.

Healy, K. 2007. Nanopore-based single-molecule DNA analysis. *Nanomedicine* 2(4), 459–481.

Hebert, P.D., Dewaard, J.R., Landry, J.-F. 2010. DNA barcodes for 1/1000 of the animal kingdom. *Biology Letters* 6, 359–362.

Hejran, A.B., Alimyar, O., Afghan, R., Niazi, P. 2024. Exploring the multifaceted role of viruses in sustaining biodiversity: Mechanisms, implications, and contributions to ecosystems stability. *Scientia Technology, Science and Society* 1(2), 35–49.

Hellens, R.P., Edwards, E.A., Leyland, N.R., Bean, S., Mullineaux, P.M. 2000. Pgreen: A versatile and flexible binary Ti vector forAgrobacterium-mediated plant transformation. *Plant Mol. Biol.* 42, 819–832.

Hipp, K., Rau, P., Schafer, B., Pfannstiel, J., Jeske, H. 2016. Translation, modification and cellular distribution of two AC4 variants of African cassava mosaic virus in yeast and their pathogenic potential in plants. *Virology* 498, 136–148.

Homs, M., Kober, S., Kepp, G., Jeske, H. 2008. Mitochondrial plasmids of sugar beet amplified via rolling circle method detected during curtovirus screening. *Virus Res.* 136, 124–129.

Horn, J., Lauster, S., Krenz, B., Kraus, J., Frischmuth, T., Jeske, H. 2011. Ambivalent effects of defective DNA in beet curly top virus-infected transgenic sugarbeet plants. *Virus Res.* 158, 169–178.

Houldcroft, C.J., Beale, M.A., Breuer, J. 2017. Clinical and biological insights from viral genome sequencing. *Nature Reviews Microbiology* 15(3), 183–192.

Iqbal, Z., Rauf, S., Hamid, I., Ahmad, S., Anjum, M.A. 2013. Diagnostic tools using DNA bar coding for the identification of pathogen races and related species: A review. *Archives of Phytopathology and Plant Protection* 46(14), 1707–1716.

Jeske, H. 2018. Barcoding of plant viruses with circular single-stranded DNA based on rolling circle amplification. *Viruses* 10(9), 469.

Jeske, H., Gotthardt, D., Kober, S. 2010. In planta cloning of geminiviral DNA: The true Sida micrantha mosaic virus. *J. Virol. Methods* 163, 301–308.

Jeske, H., Kober, S., Schafer, B., Strohmeier, S. 2014. Circomics of Cuban geminiviruses reveals the first alpha-satellite DNA in the Caribbean. *Virus Genes* 49, 312–324.

Jeske, H., Lütgemeier, M., Preiss, W. 2001. Distinct DNA forms indicate rolling circle and recombination-dependent replication of Abutilon mosaic geminivirus. *EMBO J.* 20, 6158–6167.

Joly, S., Davies, T.J., Archambault, A., Bruneau, A., Derry, A., Kembel, S.W., Wheeler, T.A. 2014. Ecology in the age of DNA barcoding: The resource, the promise and the challenges ahead. *Molecular Ecology Resources* 14(2), 221–232.

Jovel, J., Preiss, W., Jeske, H. 2007. Characterization of DNA intermediates of an arising geminivirus. *Virus Res.* 130, 63–70.

Just, K., Leke, W.N., Sattar, M.N., Luik, A., Kvarnheden, A. 2014. Detection of tomato yellow leaf curl virus in imported tomato fruit in Northern Europe. *Plant Pathol.* 63, 1454–1460.

Jyothy, A., Sabu, S.T., Dharan, S.S. 2020. Real-time PCR or quantitative PCR (QPCR)–a revolution in modern science. *World Journal of Pharmaceutical Research* 9(11), 249–264.

Karkare, S. and D. Bhatnagar. 2006. Promising nucleic acid analogs and mimics: Characteristic features and applications of PNA, LNA, and morpholino. *Applied Microbiology and Biotechnology* 71(5), 575–586.

Kasianowicz, J. J., E. Brandin, D. Branton and D. W. Deamer. 1996. Characterization of individual polynucleotide molecules using a membrane channel. *Proceedings of the National Academy of Sciences* 93(24), 13770–13773.

Katsarou, K., Bardani, E., Kallemi, P., Kalantidis, K. 2019. Viral detection: Past, present, and future. *Bioessays* 41, 1900049.

Kiryanova, O.G.Y.E., Kiselev, S.S., Panyukov, V.V.E., Nikulin, N.A., Nazipova, N.N., Chemeris, A.V., Zimin, A.A. 2024. Using multiplex virtual PCR for genetic barcoding of Tequatrovirus bacteriophages. *Bioinformatics* 19(2), 369–392.

Lambert, C., Braxton, C., Charlebois, R.L., Deyati, A., Duncan, P., La Neve, F., Khan, A.S. 2018. Considerations for optimization of high-throughput sequencing bioinformatics pipelines for virus detection. *Viruses* 10(10), 528.

Lauring, A.S., Andino, R. 2011. Exploring the fitness landscape of an RNA virus by using a universal barcode microarray. *Journal of Virology* 85, 3780–3791.

Lefeuvre, P., Martin, D.P., Elena, S.F., Shepherd, D.N., Roumagnac, P., Varsani, A. 2019. Evolution and ecology of plant viruses. *Nature Reviews Microbiology* 17, 632–644.

Lefkowitz, E.J., Dempsey, D.M., Hendrickson, R.C., Orton, R.J., Siddell, S.G., Smith, D.B. 2018. Virus taxonomy: The database of the international committee on taxonomy of viruses (ICTV). *Nucleic Acids Research* 46(D1), D708–D717.

Lewitus, E., Li, Y., Bai, H., Pham, P., Rolland, M. 2024. HIV-1 Gag, Pol, and Env diversified with limited adaptation since the 1980s. *Mbio* 15(3), e01749-23.

Liu, Y. T. 2008. A technological update of molecular diagnostics for infectious diseases. *Infectious Disorders-Drug Targets (Formerly Current Drug Targets-Infectious Disorders)* 8(3), 183–188.

Mabey, D., Peeling, R.W., Ustianowski, A., Perkins, M.D. 2004. Diagnostics for the developing world. *Nature Reviews Microbiology* 2(3), 231–240.

Maina, S., Jones, R.A. 2023. Enhancing biosecurity against virus disease threats to Australian grain crops: Current situation and future prospects. *Frontiers in Horticulture* 2, 1263604.

Manessis, G., Gelasakis, A.I., Bossis, I. 2022. Point-of-care diagnostics for farm animal diseases: From biosensors to integrated lab-on-chip devices. *Biosensors* 12, 455.

Manrao, E. A., I. M. Derrington, A. H. Laszlo, K. W. Langford, M. K. Hopper, N. Gillgren and J. H. Gundlach, J. H. 2012. Reading DNA at single-nucleotide resolution with a mutant MspA nanopore and phi29 DNA polymerase. *Nature biotechnology* 30(4), 349–353.

Martinelli, F., Scalenghe, R., Davino, S., Panno, S., Scuderi, G., Ruisi, P., Dandekar, A.M. 2015. Advanced methods of plant disease detection. A review. *Agronomy for Sustainable Development* 35, 1–25.

Marz, M., Beerenwinkel, N., Drosten, C., Fricke, M., Frishman, D., Hofacker, I.L., Töpfer, A. 2014. Challenges in RNA virus bioinformatics. *Bioinformatics* 30(13), 1793–1799.

McNally, B., A. Singer, Z. Yu, Y. Sun, Z. Weng and A. Meller. 2010. Optical recognition of converted DNA nucleotides for single-molecule DNA sequencing using nanopore arrays. *Nano letters*, 10(6), 2237–2244.

McNally, B., Wanunu, M., Meller, A. 2008. Electromechanical unzipping of individual DNA molecules using synthetic sub-2 nm pores. *Nano letters*, 8, 3418–3422.

Mehta, D., Hirsch-Hoffmann, M., Patrignani, A., Gruissem, W., Vanderschuren, H. 2017. Cider-seq: Unbiased virus enrichment and single-read, full length genome sequencing. *biorxiv* p-168724.

Milos, P. M. 2009. Emergence of single-molecule sequencing and potential for molecular diagnostic applications. *Expert review of molecular diagnostics* 9(7), 659–666.

Morilla, G., Castillo, A.G., Preiss, W., Jeske, H., Bejarano, E.R. 2006. A versatile transreplication-based system to identify cellular proteins involved in geminivirus replication. *J. Virol.* 80, 3624–3633.

Nielsen, P. E. 2010. Peptide nucleic acids (PNA) in chemical biology and drug discovery. *Chemistry & biodiversity* 7(4), 786–804.

Owor, B.E., Shepherd, D.N., Taylor, N.J., Edema, R., Monjane, A.L., Thomson, J.A., Martin, D.P., Varsani, A. 2007. Successful application of FTA classic card technology and use of bacteriophage phi 29 DNA polymerase for large-scale field sampling and cloning of complete maize streak virus genomes. *J. Virol. Methods* 140, 100–105.

Padzil, F., Mariatulqabtiah, A.R., Tan, W.S., Ho, K.L., Isa, N.M., Lau, H.Y., Abu, J., Chuang, K.P. 2021. Loop-mediated isothermal amplification (LAMP) as a promising point-of-care diagnostic strategy in avian virus research. *Animals* 12(1), 76.

Paprotka, T., Boiteux, L.S., Fonseca, M.E.N., Resende, R.O., Jeske, H., Faria, J.C., Ribeiro, S.G. 2010. Genomic diversity of sweet potato geminiviruses in a Brazilian germplasm bank. *Virus Res.* 149, 224–233.

Pereira, F., Carneiro, J., Amorim, A. 2008. Identification of species with DNA-based technology: Current progress and challenges. *Recent Patents on DNA & Gene Sequences* 2(3), 187–200.

Prassan, C., Narayan, S.B., Hillol, C., Saxena, A.K. 2021. DNA barcoding of phytopathogens for disease diagnostics and bio-surveillance. *World Journal of Microbiology and Biotechnology* 37(3).

Preiss, W., Jeske, H. 2003. Multitasking in replication is common among geminiviruses. *J. Virol.* 77, 2972–2980.

Purty, R., Chatterjee, S. 2016. DNA barcoding: An effective technique in molecular taxonomy. *Austin J Biotechnol Bioeng*. 3, 1059.

Pybus, O.G., Rambaut, A. 2009. Evolutionary analysis of the dynamics of viral infectious disease. *Nature Reviews Genetics* 10, 540–550.

Radford, A.D., Chapman, D., Dixon, L., Chantrey, J., Darby, A.C., Hall, N. 2012. Application of next-generation sequencing technologies in virology. *Journal of General Virology* 93(9), 1853–1868.

Rapireddy, S., G. He, S. Roy, B. A. Armitage and D. H. Ly. 2007. Strand invasion of mixed-sequence B-DNA by acridine-linked, γ-peptide nucleic acid (γ-PNA). *Journal of the American Chemical Society* 129(50), 15596–15600.

Richter, K.S., Gotz, M., Winter, S., Jeske, H. 2016a. The contribution of translesion synthesis polymerases on geminiviral replication. *Virology* 488, 137–148.

Richter, K.S., Serra, H., White, C.I., Jeske, H. 2016b. The recombination mediator rad51D promotes geminiviral infection. *Virology* 493, 113–127.

Rodenburg, C. M., Y. Li, S. A. Trask, Y. Chen, J. Decker, D. L. Robertson and F. Gao. 2001. Near full-length clones and reference sequences for subtype C isolates of HIV type 1 from three different continents. *AIDS research and human retroviruses* 17(2), 161–168.

Rodríguez, R.A., Pepper, I.L., Gerba, C.P. 2009. Application of PCR-based methods to assess the infectivity of enteric viruses in environmental samples. *Applied and Environmental Microbiology* 75(2), 297–307.

Romay, G., Chirinos, D., Geraud-Pouey, F., Desbiez, C. 2010. Association of an atypical alphasatellite with a bipartite new world begomovirus. *Arch. Virol.* 155, 1843–1847.

Roossinck, M.J. 2012. Plant virus metagenomics: Biodiversity and ecology. *Annual Review of Genetics* 46(1), 359–369.

Rosenstein, J. K., M. Wanunu, C. A. Merchant, M. Drndic and K. L. Shepard. 2012. Integrated nanopore sensing platform with sub-microsecond temporal resolution. *Nature methods* 9(5), 487–492.

Sahu, B., I. Sacui, S. Rapireddy, K. J. Zanotti, R. Bahal, B. A. Armitage and D. H. Ly. 2011. Synthesis and characterization of conformationally preorganized,(R)-diethylene glycol-containing γ-peptide nucleic acids with superior hybridization properties and water solubility. *The Journal of organic chemistry* 76(14), 5614–5627.

Shendure, J., R. D. Mitra, C. Varma and G. M. Church. 2004. Advanced sequencing technologies: methods and goals. *Nature Reviews Genetics* 5(5), 335–344.

Shendure, J., G. J. Porreca, N. B. Reppas, X. Lin, J. P. McCutcheon, A. M. Rosenbaum and G. M. Church. 2005. Accurate multiplex polony sequencing of an evolved bacterial genome. *Science* 309(5741), 1728–1732.

Shepherd, D.N., Martin, D.P., Lefeuvre, P., Monjane, A.L., Owor, B.E., Rybicki, E.P., Varsani, A. 2008. A protocol for the rapid isolation of full geminivirus genomes from dried plant tissue. *J. Virol. Methods* 149, 97–102.

Shestopalov, A.M., Kononova, Y.V., Gadzhiev, A.A., Gulyaeva, M.A., Vasfi, M.M., Alekseev, A.Y., Jamalutdinov, J.M., Shchelkanov, M.Y., 2020. Biodiversity and epidemic potential of Chiropteran coronaviruses (Nidovirales: Coronaviridae). *South of Russia: Ecology, Development* 15(2), 17–34.

Siddell, S.G., Smith, D.B., Adriaenssens, E., Alfenas-Zerbini, P., Dutilh, B.E., Garcia, M.L., Junglen, S., Krupovic, M., Kuhn, J.H., Lambert, A.J. 2023. Virus taxonomy and the role of the international committee on taxonomy of viruses (ICTV). *Journal of General Virology* 104, 001840.

Silva, S.J.R.D., Pardee, K., Pena, L. 2019. Loop-mediated isothermal amplification (LAMP) for the diagnosis of Zika virus: A review. *Viruses* 12(1), 19.

Simmonds, P., Adams, M.J., Benkő, M., Breitbart, M., Brister, J.R., Carstens, E.B., Davison, A.J., Delwart, E., Gorbalenya, A.E., Harrach, B., Hull, R. 2017. Virus taxonomy in the age of metagenomics. *Nature Reviews Microbiology* 15(3), 161–168.

Singer, A., S. Rapireddy, D. H. Ly and A. Meller. 2012. Electronic barcoding of a viral gene at the single-molecule level. *Nano Letters* 12(3), 1722–1728.

Singer, A., M. Wanunu, W. Morrison, H. Kuhn, M. Frank-Kamenetskii and Meller, A. 2010. Nanopore-based sequence-specific detection of duplex DNA for genomic profiling. *Biophysical Journal* 98(3), 598a–599a.

Singh, C., Roy-Chowdhuri, S. 2016. Quantitative real-time PCR: Recent advances. *Clinical Applications of PCR* 161–176.

Soni, G. V., A. Singer, Z. Yu, Y. Sun, B. McNally and A. Meller. 2010. Synchronous optical and electrical detection of biomolecules traversing through solid-state nanopores. *Review of Scientific Instruments*, 81(1), 014301.

Stenger, D.C., Revington, G.N., Stevenson, M.C., Bisaro, D.M. 1991. Replicational release of geminivirus genomes from tandemly repeated copies: Evidence for rolling-circle replication of a plant viral DNA. *Proc. Natl. Acad. Sci. USA* 88, 8029–8033.

Tang, Y.W., Procop, G.W., Persing, D.H. 1997. Molecular diagnostics of infectious diseases. *Clinical Chemistry* 43(11), 2021–2038.

Tatineni, S., Hein, G.L. 2023. Plant viruses of agricultural importance: Current and future perspectives of virus disease management strategies. *Phytopathology®* 113, 117–141.

Thielecke, L., Aranyossy, T., Dahl, A., Tiwari, R., Roeder, I., Geiger, H., Cornils, K. 2017. Limitations and challenges of genetic barcode quantification. *Scientific Reports* 7(1), 43249.

Thompson, M. A., J. A Aberg, P. Cahn, J. S. Montaner, G. Rizzardini, A. Telenti and R.T. Schooley. 2010. Antiretroviral treatment of adult HIV infection: 2010 recommendations of the International AIDS Society -USA panel. *Jama* 304(3), 321–333.

Trovão, N.S., Baele, G., Vrancken, B., Bielejec, F., Suchard, M.A., Fargette, D., Lemey, P. 2015. Host ecology determines the dispersal patterns of a plant virus. *Virus Evolution* 1(1), vev016.

Vernet, G. 2004. Molecular diagnostics in virology. *Journal of Clinical Virology* 31(4), 239–247.

Wang, X., Liu, C., Tan, Z., Zhang, J., Wang, R., Wang, Y., Jiang, X., Wu, B., 2023. Population genetics and phylogeography of alfalfa mosaic virus in China and a comparison with other regional epidemics based on the cp gene. *Frontiers in Plant Science* 13, 1105198.

Wang, Y., Tang, C.Y., Wan, X.F. 2022. Antigenic characterization of influenza and SARS-CoV-2 viruses. *Analytical and Bioanalytical Chemistry* 414(9), 2841–2881.

Wanunu, M., W. Morrison, Y. Rabin, A. Y. Grosberg and A. Meller. 2009. Electrostatic focusing of unlabelled DNA into nanoscale pores using a salt gradient. *Nature nanotechnology* 5(2), 160–165.

Wanunu, M., J. Sutin, B. McNally, A. Chow and Meller, A. 2008. DNA translocation governed by interactions with solid-state nanopores. *Biophysical journal* 95(10), 4716–4725.

Wyant, P.S., Gotthardt, D., Schafer, B., Krenz, B., Jeske, H. 2011. The genomes of four novel begomoviruses and a new Sida micrantha mosaic virus strain from Bolivian weeds. *Arch. Virol.* 156, 347–352.

Wyant, P.S., Kober, S., Schwierzok, A., Kocher, C., Schafer, B., Jeske, H., Wege, C. 2012a. Cloned tomato golden mosaic virus back in tomatoes. *Virus Res.* 167, 397–403.

Wyant, P.S., Strohmeier, S., Schafer, B., Krenz, B., Assuncao, I.P., Lima, G.S.D., Jeske, H. 2012b. Circular DNA genomics (circomics) exemplified for geminiviruses in bean crops and weeds of Northeastern Brazil. *Virology* 427, 151–157.

Zhang, N. and D. H. Appella. 2010. Advantages of peptide nucleic acids as diagnostic platforms for detection of nucleic acids in resource-limited settings. *Journal of Infectious Diseases* 201, S42–S45.

Zuo, K., Gao, W., Wu, Z., Zhang, L., Wang, J., Yuan, X., Liu, H. 2024. Evolution of virology: Science history through milestones and technological advancements. *Viruses* 16(3), 374.

5 Mapping Bacterial Biodiversity Using DNA Barcoding

Amjad Abbas, Muhammad Usman, Muhammad Huzaifa Tanveer, Aysha Khalid, Muhammad Ehetisham-Ul-Haq, Amer Habib, Muhammad Zulkiffal, and Muhammad Amjad Ali

5.1 INTRODUCTION

5.1.1 Overview of Bacterial Biodiversity

Taxonomic studies reveal diverse bacterial classes with varying abilities and roles in shaping or degrading other living entities. Such diversity is supported by the presence of bacteria in extreme environments, including glaciers, aquatic systems, and hot springs. For example, certain bacterial species, such as *Rhizobia* and *Azotobacter*, are crucial for most plants, as they convert environmental nitrogen into a usable form. Such involvement of bacteria strengthens the ecosystem, enabling it to achieve a sustainable food supply from the bottom to the top of the food web. In aquatic and marine ecosystems, oxygen availability is crucial for the survival of animals and plants. In such a scenario, bacteria play a crucial role in providing oxygen, thereby controlling aquatic life. Cyanobacteria, in this regard, are considered the lungs of marine life. Bacteria possess an extraordinary ability to survive, with rapid adaptability to changing environments and harsh conditions (Stal, 2007).

Due to their widespread presence, bacteria exhibit tremendous physical and genetic diversity. The diversity of bacteria has been studied, beginning with simple culturing techniques, biochemical characterization, and advancing to sophisticated strategies based on DNA sequencing and barcoding methods (Pontes et al., 2007). Although culturing bacteria provides ample information regarding their diversity, when carried out with available differentials, many species of bacteria still cannot be cultured. Furthermore, even some culturable bacteria possess similar properties, as determined by biochemical approaches. Such hurdles make it difficult to identify closely related species. Bacterial species with great diversity are present in the human gut and contribute to digestion by providing vitamins and other essential elements that are not possible without these tiny creatures. However, bacterial species can also lead to indigestion and gastric diseases. Such pathogenic bacteria are not limited to humans, as they also cause diseases in other forms, which impose threats to agricultural produce both in the

 DOI: 10.1201/9781003538929-5

field and after harvesting. Diversity is a challenging subject that helps us understand the effects of climate change on bacteria, and it also aids in taking measures to manage plant diseases and preserve precious food commodities. Several techniques are employed to identify the differences and similarities between bacteria, enabling a valid approach to disease mitigation (Varadi et al., 2017).

In crop fields, several factors favor the growth and abundance of specific microbial communities. These factors that are important for the promotion of some bacterial species over others should be recognized as basic factors. The type of soil structure and pH are crucial in determining which soil microbiota will dominate (Islam et al., 2020). Very loamy soils have a low amount of available oxygen, and the diversity of bacteria is relatively low compared to soils that have sufficient pores and an abundance of oxygen. The small size of particles and compactness in loamy soil support a lesser number of bacterial species. Soils also differ in pH when comparing two areas. Very acidic soils do not support microbial communities as effectively as basic soil types (Ren et al., 2018). Organic matter is important for the subsistence of microbes in the soil. Soil rich in organic matter is considered a rich source of microbial diversity compared to areas with soils lacking organic matter. Typically, areas receiving sufficient rainfall with a wide distribution, such as regions with a tropical climate, exhibit a greater diversity of bacteria (Mohammadi et al., 2011).

Nitrogen fixation is a crucial function of soil-inhabiting bacteria, making nutrients available in a usable form. In the soil, there is also competition among bacteria for nutrition, symbiosis, and survival. These factors also determine possible diversity of bacteria in the soil. The diversity is not limited to friendly bacteria, as many soil bacteria are pathogenic and can infect plants (Tilak et al., 2005). For instance, the soil pathogenic bacterium *Xanthomonas* sp. causes citrus canker disease. Citrus orchards, being a major fruit crop, are inhabited by a diverse range of pathogenic bacteria that become more aggressive in ever-changing climatic conditions. The diversity of pathogenic bacteria is more important for disease management in plants, while beneficial bacteria are studied to increase their population for more beneficial effects (Naqvi et al., 2022).

Bacterial populations exist in the soil and within plants, serving either as growth promoters or as potent plant pathogens. Bacteria also contribute toward maintaining an ecosystem's balance. As the ecological conditions in the ecosystem shift, the bacterial diversity and its activities also shift. Since the ecosystems of forests, grasslands, and deserts are diversified, similarly, the diversity of bacterial communities within these ecosystems is also observed.

For instance, *Acidobacteria* dominate extremely acidic soils, whereas *Actinobacteria* and *Bacteroidetes* are more commonly found in soils with higher pH levels. The study also brings into focus the functionality of bacteria that inhabit heterogeneously in the living conditions, which possess varied capabilities toward recycling nutrients in soil, such as nitrogen, carbon, and phosphorus, and so on (Kalam et al., 2020)

Soil pH, organic matter, moisture, and texture are significant factors in determining which bacterial populations are likely to be dominant in a given habitat. Soil structures of various types facilitate the growth of specific bacterial populations by regulating the supply of oxygen and nutrients in the soil. Similarly, bacteria present in areas with drought comprise a genetic background that supports survival against water constraints. Abiotic and biotic factors influence the diversity of soil bacteria. This research illustrates the interaction between bacterial diversity in the environment and vice versa, and it establishes the concept clearly (2018). Bacteria present in the soil also assist in the fixation of nitrogen and its availability to plants. Bacterial populations also exhibit competitive or symbiotic responses that enhance their survival. Due to these facts, the diversity of bacteria is significant in the ecosystem (Hayat et al., 2010). Pathogenic bacteria, such as *Xanthomonas* sp., which cause citrus canker, are substantial in terms of disease management, as climatic changes bring about diversity.

5.1.2 Bacterial Characterization

5.1.2.1 Traditional Methods of Bacterial Characterization

5.1.2.1.1 Morphological Characterization

Bacteria are identified based on various morphological features. These features are helpful in diversity studies. Morphological determinants, such as growth conditions, type of growth medium, colony shape, colony size, growth rate, growth at a particular temperature, response to a specific kind of medium, and many other features, are considered when studying the morphological diversity of bacteria. These observations have been made since the inception of microbiology studies and are still considered the main criterion for identifying bacteria and their functional profiling (Sousa et al., 2013).

The structure and composition of various organelles or their constituents are helpful to differentiate bacterial species. Gram staining targets the peptidoglycan layer in the cell wall of bacteria. Gram-positive bacteria have the ability, due to the presence of peptidoglycan, to retain the effects of dye, while gram-negative bacteria don't take up dye. This technique provides a basic classification of the bacteria, which is aided by their pathogenicity and becomes a good selection marker (Thanassi et al., 2012). For example, *Escherichia coli* is a gram-negative, rod-shaped bacterium, while *Bacillus subtilis* is a Gram-positive bacterium with a thick peptidoglycan layer of cell wall. *Pseudomonas aeruginosa*, with a thin layer, does not respond to crystal violet because it is a Gram-negative bacterium (Suvorov et al., 2008).

Other dyes, such as acid-fast staining, are a particular test to identify *Mycobacterium*, and endospore staining for *Bacillus* sp. is a unique test for identification based on morphological grounds. Microscopic features of bacteria also enable morphological characterization at the detail level with the light microscope, electron microscope, or fluorescence microscope. Microscopy separates bacteria based on cell shape and presence or absence of pili and flagella (Beveridge, 2001). Other tests, such as the KOH assay and microscopy, along with infection tests, have evolved over time to characterize the bacteria and determine their pathogenic fate. Such classical methods are easy to identify bacterial species and have been summarized in Figure 5.1.

On the other hand, there are several drawbacks to the morphological identification of bacteria, as many bacterial species exhibit similar color patterns and identical shapes, making it challenging to distinguish them from other species. For instance, *E. coli* and *Shigella* spp. are morphologically identical on the basis of growth pattern, which results in confusion to easily differentiate these bacteria (Tshikhudo et al.,

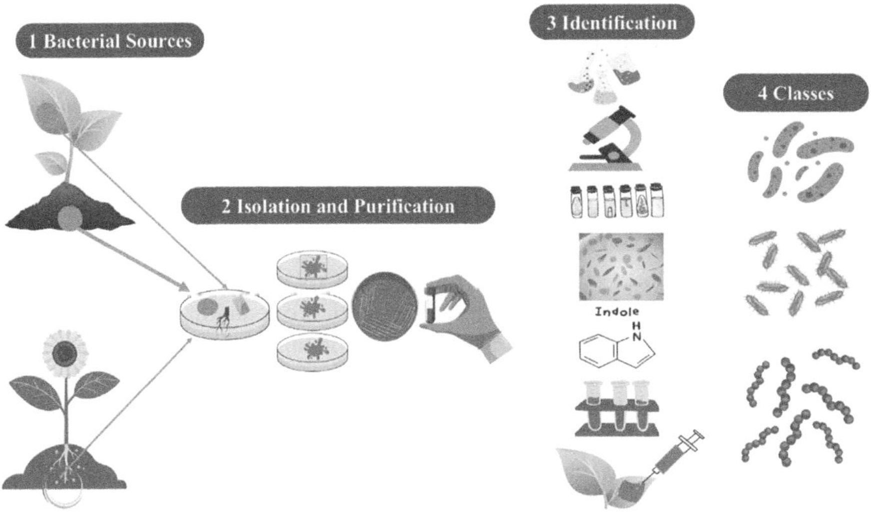

FIGURE 5.1 The scope of traditional identification layout for the classification of pathogenic and non-pathogenic bacteria.

2013). As mentioned above, certain species are unculturable in the lab; therefore, for those, traditional morphological characterization is not feasible, and this results in the conclusion that unculturable bacterial species in the laboratory are difficult to identify using these methods (Stewart, 2012).

The usual differences in taxonomy among different microbes, based on biochemical tests alone, are not sufficient to segregate between two species (Vandamme et al., 1996). Molecular techniques are alternative methods for better resolution of the solution to this problem. Initially, 16S fragments of the ribosomal gene were considered as sufficient markers for identifying the bacteria with this fine filter. With the advancement in understanding DNA sequencing, other ubiquitous genes have been included to strengthen the concept of taxonomy (Fukuda et al., 2016; Maurer, 2011).

5.1.2.1.2 Biochemical and Physiological Characterization

Biochemical and physiological characterization is used for assessing growth potential, in vitro antibiotic assays, taxonomic positioning, and to place the pathogen in a particular position. Bacteria are identified through various biochemical traits as a profiling tool for classification purposes. For example, *P. aeruginosa* is determined by the pyocyanin production and glucose oxidation (Alatraktchi et al., 2020).

Physiological experiments with the growth of bacterial cultures are tested under varied conditions, using diverse ranges of factors such as pH, temperature, nutritional quality, and salinity, which help to identify suitable bacterial strains. For instance, *Thermus aquaticus* can survive in high temperatures, and *Halobacterium salinarum* also thrives in high-salinity conditions (Ranawat and Rawat, 2017). Like biochemical tests, this technique also has limitations. Closely related species such as *B. subtilis* and *B. licheniformis* possess the same metabolic profiles, making it hard to differentiate these two species on this basis (Voigt et al., 2004). Biochemical profiling of bacteria is commonly used in clinical and environmental microbiology due to its ease of use and the availability of inexpensive sources for characterization purposes (Rave et al., 2018).

5.1.2.2 Importance of Characterizing Bacteria

5.1.2.2.1 Function in Grasping Bacterial Diversity and Ecosystem Function

Characterization of bacteria is a key component in knowing the diversity and activities of bacteria in the ecosystem, e.g., bacteria that are found in the soil are useful in the process of nutrient recycling in the soil, decomposition of organic matter, and make them accessible to the other organisms, and also useful in the maintenance of plant health as plants utilize them in their growth and other metabolic activities (Khatoon et al., 2017). These traditional methods are significant because they provide preliminary and fundamental knowledge on the isolation of nitrogen-fixing bacteria, such as *Azotobacter* and *Rhizobium* (Kawaka, 2022). The same methods are not universal and cannot be applied to describe the bacteria of all origins because complex habitats, such as aquatic and soil ecosystems, are quite different (Pontes et al., 2007).

DNA barcoding is a new and ancillary tool for bacterial identification. The method can be applied to both culturable and non-culturable bacterial species for identification, aiding in the determination of the biological function of various bacterial communities (Fakruddin and Mannan, 2013). For example, discovery of new bacterial species through the use of metagenomics analyses on the occurrence of nitrogen-fixing bacteria in soils demonstrates the synergistic value that can exist between conventional and molecular techniques to identify bacterial diversity and bacterial species (Rajguru et al., 2024).

5.1.2.2.2 Applications in Environmental, Medical, and Industrial Microbiology

The use of bacterial characterization is highly significant and in high demand in numerous diverse disciplines. In environmental microbiology, it is applied for tracking populations of bacteria or other microbes present in the soil, water, and air of the ecosystems. Identification of enhanced bioremediation processes through characterization of bacterial populations available at the wastewater treatment plants (Bouchez et al., 2016)

The application of phenotypic methods is critical for comprehending bacteria in terms of functional characterization and identifying their disease-causing role in medical microbiology. Since bacteria are rapidly acquiring and developing antibiotic resistance, there is a need for more precise and rapid identification techniques, as DNA barcoding plays this function more accurately. In industrial microbiology, for optimizing processes such as fermentation and biofuel production, characterizing bacterial cultures is of great significance for food safety. For example, the lactic acid fermentation process, where the bacteria play the major role, is significant to study prior to use in food and pharmaceutical industries. The characterization of lactic acid bacteria is very useful for industry to select the right strain for more efficient fermentation of food with enhanced quality and safety.

5.1.3 Bacterial Identification

5.1.3.1 Classical Methods of Bacterial Identification

5.1.3.1.1 Bacterial Characterization via Selective Media and Phenotypic Traits

Various media are used for the growth and isolation of beneficial and disease-causing bacteria. Differential and selective media, such as Mannitol Salt Agar for the *Staphylococcus* spp., and MacConkey Agar media for Gram-negative bacteria, have critical applications in the clinical microbiology laboratory for the isolation and identification of pathogens such as *Salmonella* and *E. coli*, according to the metabolic characteristics and growth on the selective media, that is, lactose or mannitol fermentation (Wadt, 2024).

Such physiological markers are also equally important in certain circumstances. For instance, phenotypic identification depends on metabolic activities, morphological growth, endospore formation, morphology, and anaerobic activity in *Clostridium* spp., but lacks specificity at the level of closely related species (Al-Hinai et al., 2015).

5.1.3.2 Shift toward Molecular Techniques

5.1.3.2.1 Shift to Genotypic Identification and DNA Barcoding

Genetic markers are now utilized for the genotypic identification of bacterial species and are also applicable to unculturable bacteria, based on 16S rRNA, which is the most prevalent procedure for identification (Church et al., 2020). With the advent of molecular tools, such as PCR, markers like RAPD, SSR, RFLP, and other DNA-based markers are used for the identification of any bacterial species (Amom and Nongdam, 2017). Despite this, there are restrictions with these methods too because of greater similarity index between close-related species, which makes it difficult to distinguish them independently, as the presence of reference sequences makes it more complex within the same taxa, as they are primarily presented with the same restriction pattern (Harari et al., 2010).

Since DNA barcoding resolves these challenges, it requires less time and provides precise results using universal genetic markers and high-throughput sequencing (Mishra et al., 2016). This DNA barcoding method facilitates the identification of bacterial diversity on an enormous scale, as it is not based on culturing the pathogen and prior information about sequencing (Amom and Nongdam, 2017). Most frequently applied DNA barcodes, such as 16S rRNA, offer fine insight into microbial diversity, providing a deeper understanding of the biogeochemical processes that occur in ecosystems by changing the manner of studying microbial ecology (Joly et al., 2014).

5.1.4 Introduction to DNA Barcoding

The DNA barcoding method was initially developed and introduced when specific and standardized genetic markers were employed for the identification and categorization of species into various groups. This method has greatly revolutionized research studies in biodiversity. Despite this, the method was created for identifying animal breeds using genetic markers and gene sequences, which consider variable and conserved regions to distinguish species by applying universal primers. This method is important for achieving an understanding of microbial diversity in marine and agricultural environments (Joly et al., 2014).

DNA barcoding was highly effective in identifying and examining bacterial diversity, predominantly in seafloor and seawater, as it demonstrated the structure of beta-diversity and showed that these bacteria act as carbon and nitrogen co-fixers to ensure nutrient balance in the environment. Another study has indicated that DNA barcoding also assists in identifying bacterial diversity in soil microbiology (Earth, 2018). It revealed that the biotic components, such as plant root exudates, and the abiotic components, such as soil pH and texture, play a crucial role in determining the bacterial communities in the soil because these bacterial communities are beneficial for nitrogen fixation and supply it to the plants, and this could be achieved only by DNA barcoding because it helped us to determine the bacterial species present in closely related communities (Adeniji et al., 2024).

DNA barcoding is also beneficial in clinical microbiology, as it discriminates against bacteria that are resistant to antibiotics by detecting specific barcode regions, such as those of *E. coli* and *S. aureus*. It is beneficial in diagnostic interventions and assisting in the treatment of the disease. Likewise, DNA barcoding is also helpful in various fields, such as industrial microbiology, as it is applied to optimize processes where bacteria are utilized, including the bioremediation process. It is used in the monitoring process, where bacteria break down heavy metals and hydrocarbons, providing improved insight into the bacterial community in the treatment of environmental hazardous materials and contaminants (Li et al., 2020).

Additionally, there are certain limitations in species discrimination, similar to that of 16S rRNA, based on the fact that some markers cannot distinguish between closely related species. Other markers, equally important in bacterial identification, include Internal Transcribed Spacer (ITS) and housekeeping genes such as *gyrB* and *rpoB*. Improved databases and established protocols, such as Greengenes and SILVA, are commonly used for enhanced bacterial community understanding at the taxonomic resolution, as well as for appropriate classification (Volokhov et al., 2012).

DNA barcoding is an accepted method of sequencing that is in an advanced phase, and applications of bioinformatics tools for DNA barcoding are useful to determine specific bacterial communities. In addition, the application of Next Generation Sequencing for microbial communities enables high-throughput analysis, while the application of metagenomics provides deeper analysis of bacterial functions by combining the DNA barcoding strategy with whole-genome sequencing. The application of powerful bioinformatics tools for DNA barcoding, such as MOTHUR and QIIME, simplifies sequencing analysis and measures biodiversity metrics with high precision and accuracy (Wani et al., 2021).

5.2 DNA BARCODING TECHNIQUES

5.2.1 Selection of Barcode Regions

Barcode region selection is one of the key and most critical steps in identifying bacterial cultures for accurate DNA barcoding, as it influences the resolution of species. The most widely used DNA barcode region is the 16S rRNA, as it is universally present, has variable regions for discriminating among various species, and conserved regions for universal primers. The 16S rRNA fragment is not ideal for distinguishing closely related species like *E. coli* and *Shigella*. Therefore, so many other barcode regions are of higher resolution than 16S rRNA. Various alternative barcode regions enhance our reliability in identifying bacteria (Antil et al., 2023).

1. **Internal Transcribed Spacer (ITS)**:
   The Internal Transcribed Spacer is well-suited for the identification of closely related species, as it is more variable than 16S rRNA. However, it also has its pitfalls, with alignment and primer design being key concerns or obstacles in employing this method (Cheng et al., 2014).
2. **Housekeeping Genes**:
   Some genes are present in many bacterial species for DNA barcoding. The genes include *rpoB*, *dnaK*, *recA*, *gyrB*, *fusA*, and *rpoD*. These genes facilitate the accurate identification and provide improved resolution. For example, the *rpoB* gene provides better resolution in *Mycobacterium*, *dnaK* in *Bacillus*, *recA* in *Rhizobium*, *gyrB* in *Pseudomonas*, *fusA* in *Salmonella*, and *rpoD* in *Escherichia*. These genes are all species-specific for DNA barcoding and are used for the identification of bacteria (Ramesh et al., 2021).
3. **23S RNA Gene:**
   This gene is used for the identification of bacterial species belonging to the same group, such as *S. aureus* and *S. epidermidis*. It offers precise variations in the DNA barcoding. However, this system is not widely used because it cannot distinguish between two species with a close genetic relationship (Krawczyk and Kur, 2018).
4. **Cpn60 Gene:**
   The Cpn60 encodes a transcript that involves the folding process of other proteins. These are folding chaperonin with a central role in the folding and structure determination of different proteins in the cell. These are used as molecular signatures for bacterial identification (Lund, 2009).
5. **Multi-Locus Sequence Typing (MLST):**
   For more precise identification and taxonomy determination, many housekeeping genes are sequenced with universal primers and analyzed. This approach is more reliable for speciation studies of bacteria and is often applied in DNA barcoding for better resolution (Maiden, 2006).

5.2.1.1 Selection Criteria

The choice of barcode regions is based on the sample type, study objectives, and taxonomic resolution. If the sample type is from a diverse habitat, then 16S rRNA is applied for DNA barcoding, and for the sample type of clinical significance, housekeeping gene or ITS *are* applied for DNA barcoding because they yield information of more accurate taxonomy (Church et al., 2020)

5.2.2 Sequencing Methods

In DNA barcoding, the most important step is sequencing, as it is used to check the amplified DNA barcode region for the identification of the bacteria. Sanger sequencing remains the gold standard for small-scale bacterial identification studies, as it is highly reliable due to its high accuracy. However, it has limited throughput and is increasingly cost-effective. When examining large-scale biodiversity, species screening may ultimately be used in only a very small number of circumstances (Sanger et al., 1977). With the advent of Next-Generation Sequencing (NGS) methods, such as those from PacBio and Illumina, the world of DNA barcoding has been revolutionized by high-throughput sequencing capacity. Illumina methods are particularly suitable for the sequencing of 16S rRNA amplicons, as they provide generationally high sequence coverage useful for quantifying marker gene diversity across a microbial community. On the other hand, PacBio's long reads lend themselves to enhancing taxon resolution for closely related species; however, higher error rates are usually experienced and typically require stringent quality-control measures. Ion Torrent provides low-cost, fast sequencing ideal for biodiversity studies, and Oxford Nanopore, in addition to being low-cost, also provides a real-time sequencing capability; however, it does require strict error correction (Bartram et al., 2011).

5.3 DATA ANALYSIS AND INTERPRETATION

5.3.1 Bioinformatics Tools and Pipelines

Bioinformatics tools are essential for analyzing and processing DNA barcoding data, and several such resources are available. QIIME includes specialized tools for filtering, clustering, and diversity analysis of 16S rRNA sequences. MOTHUR also features user-friendly modules for sequence alignment, chimera detection, and diversity analysis, which generate a useful taxonomic resolution. Another Bioconductor tool, DADA2, addresses the exact sequence variants for exploring high-resolution amplicon sequence data, independently of OTU-based or other traditional methods. Both USEARCH and VSEARCH offer quick-to-use sequence clustering and taxonomic classification of sequence data, with VSEARCH representing an open-source version of a search tool. PICRUSt predicts likely functional profiles of a microbial community from 16S rRNA sequence data produced by high-throughput sequencing. LEfSe and MetaPhlAn can identify biomarkers from sequence data and/or provide profiles and statistical measurements for metagenomic communities (Fjukstad and Bongo, 2017). All of the techniques mentioned earlier play a central role in bacterial identification, and this concept is summarized in Figure 5.2.

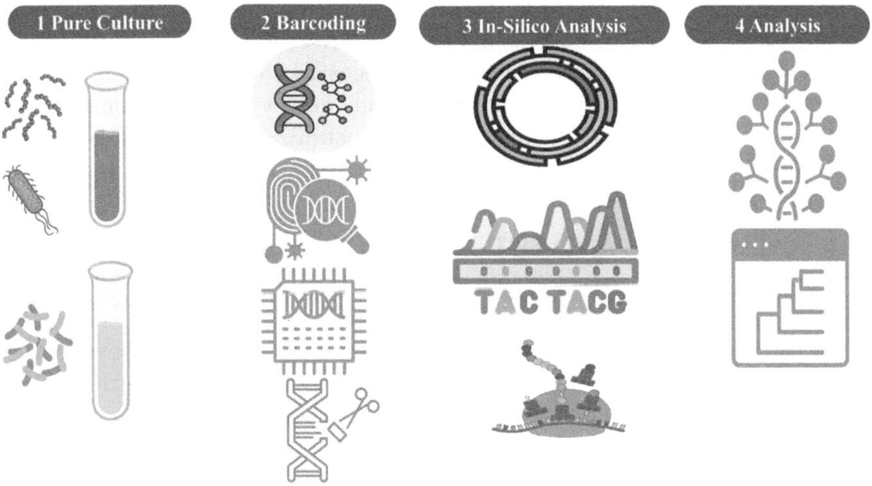

FIGURE 5.2 The process of identification of bacteria with modern tools, which involves pure culture and DNA fingerprinting based on ***in silico*** tools to establish a phylogenetic tree of the closely related species.

5.3.2 Biodiversity Metrics

Biodiversity metrics provide a means to quantify microbial community structure and function. Species richness counts the number of unique bacterial species and, therefore, provides a measure of the health of an ecosystem. Species evenness measures how abundance is distributed, which indicates a level of functional stability. Shannon and Simpson indices take richness or evenness measures and combine them into a single metric, or, as in Simpson's case, measure dominance. Chao1 may be used to measure richness in rare species. Phylogenetic diversity is a measure of evolutionary relatedness, and beta diversity measures dissimilarity in communities. Functional diversity goes beyond taxon to assess whether there are functional roles. Pielou's evenness and Jaccard's similarity measure evenness and similarity in communities (Pavoine and Bonsall, 2011).

5.3.2.1 Applications of DNA Barcoding in Agricultural Microbiology

DNA barcoding plays a pivotal role in agricultural and industrial microbiology by identifying microbes that contribute to soil fertility and the biodegradation of contaminants. In agriculture, crop rotation and sustainable practices, such as organic farming, increase the diversity of microbial communities that help recycle nutrients, while also increasing crop yields. For bioremediation, DNA barcoding enables the identification of bacteria that recycle hydrocarbons, heavy metals, and pesticides in contaminated soils, thereby contributing to an ongoing, sustainable cleanup (Hafez and Elbestawy, 2009). For industrial microbiology, DNA barcoding has enabled the identification of essential bacterial strains for fermented products, including cheese, food products, and wines. It has applications for flavor optimization and standardizing quality. DNA barcoding is used to identify bacteria known to contribute to biofuel production. Bacteria that enhance bioethanol and biogas yields have been identified, and anaerobic microbial communities have been optimized for pollutant degradation and sustainability in wastewater treatment plants (Choudhary et al., 2021).

5.4 CHALLENGES AND FUTURE DIRECTIONS

5.4.1 Technical and Methodological Challenges

DNA barcoding can be informative for assessing bacterial diversity; however, several technical and methodological limitations or biases also exist, which can limit its accuracy and utility. There are many types of bacterial barcodes; the 16S rRNA gene is the primary type used and typically exhibits variability that is insufficient to differentiate between closely related species. When similar species cannot be differentiated, a false identification results, and the taxonomy of the community may not be completely resolved, especially in complex microbial communities. Furthermore, due to various methods of DNA extraction and PCR, the representation of microbial communities may be compromised. Some DNA extraction methods preferentially lyse certain bacteria, while certain PCR methods amplify specific sequences; therefore, the representation of a community could be biased. Measurement of microbial diversity in certain environments, such as soil or marine habitats, is challenging to predict, and sometimes protocol optimization is required to accurately present microbial diversity. Additional challenges include incomplete reference databases and high technical requirements, which also pose hurdles to the implementation of protocols. Many bacterial species, especially uncultured ones, are not included in databases, which restricts taxonomic classification, and reference databases can result in certain errors.

The costs of sequencing, alongside technical understanding, bioinformatics tools, and computational resources, restrict accessibility. In addition, DNA extraction from low-biomass samples, such as air or water, is also limited by the prospect of contamination and lower yield. Furthermore, distinguishing live cells from dead cells can result in overestimating diversity. Diversity assessments and rare species detection in highly diverse communities must be interpreted with caution due to the underrepresentation of taxa in sequencing data, which renders it impossible to accurately assess overall biodiversity (Sheth and Thaker, 2017).

5.4.2 Advancements in DNA Barcoding

Technological advancements in DNA-based barcoding have significantly improved its accuracy and utility in microbial diversity studies. The selection of different barcode regions, such as the Internal Transcribed Spacer (ITS) and housekeeping genes (rpoB, gyrB), provides greater taxonomic resolution than the 16S rRNA gene, allowing the ability to separate closely related species, for instance, in a complex bacterial community. Moreover, with the advent of next-generation sequencing (NGS) platforms, such as Illumina and PacBio, high-throughput sequencing is enabling barcoding to be carried out with improved coverage and depth (Franco et al., 2021). For example, PacBio creates long reads, producing full-length 16S rRNA sequences. Although these long reads provide improved mean accuracy in taxonomy, they have a larger error profile that requires additional error correction. These advancements will enable research projects to investigate microbial communities in greater detail than previous projects (Poirier et al., 2018).

5.4.3 Emerging Technologies in DNA Barcoding

5.4.3.1 Third-Generation Sequencing (PacBio, Nanopore)

Advancements in third-generation sequencing (TGS) technologies, particularly those from PacBio (Pacific Biosciences) and Oxford Nanopore, are enhancing DNA barcoding and microbial ecology, allowing for greater resolution. PacBio made its single-molecule real-time (SMRT) sequencing technology public in 2009. PacBio SMRT sequencing can produce long read lengths (10–25 kb) with greater than 99% accuracy through circular consensus sequencing, allowing for full-length sequencing of the 16S rRNA gene (~1.5 kb) at the nucleotide level. This is especially valuable because the increased resolution can distinguish closely related species, such as *Escherichia* spp. and *Shigella* spp., which are important for biodiversity studies in complex environments, particularly in soil. Furthermore, the development of hybrid sequencing, which combines PacBio long-reads with Illumina short-reads, effectively generates complete bacterial genome assemblies, fully resolving complex genomic regions, such as ribosomal operons (Kim et al., 2010).

The Oxford Nanopore MinION sequencing platform offers ultra-long sequencing reads (>100 kb) and real-time, portable sequencing capabilities, thereby greatly reducing PCR biases by sequencing native DNA. Current uses in clinical microbiology underscore the translational value of TGS. These technologies, when applied to monitor the spread of antibiotic resistance genes in hospital settings, illustrated their value in real-time monitoring of microbial community evolution following therapeutic interventions. Rapid sequencing of whole bacterial genomes has also streamlined outbreak investigations, enabling the accurate tracing of pathogen transmission routes.

Despite these benefits, the cost and complexity of data analysis remain issues. These issues are being addressed quickly through improvements in sequencing chemistry and bioinformatics tools, making TGS more accessible for everyday use in biodiversity research. As such technologies advance, they have the potential to expose even deeper levels of microbial diversity and function in a wide range of ecosystems (Oehler et al., 2023).

5.4.3.1.1 Single-Cell Genomics

Single-cell genomics (SCG) has proven to be a revolutionary method for studying microbial biodiversity, especially for bacteria that do not cultivate in the lab. Classical metagenomic methods pool signals from entire microbial communities, thereby obscuring genetic and functional heterogeneity at the single-cell level. SCG compensates for this deficiency by allowing the genomic survey of individual microbial cells, providing biodiversity studies with unprecedented resolution. The basis of contemporary SCG is rooted in sophisticated cell isolation methods. Fluorescence-activated cell sorting (FACS) enables the specific selection of single bacterial cells based on features, including size, shape, or fluorescence labeling.

Single-cell genomics (SCG) has revealed that CPR bacteria possess small genomes and engage in metabolic interactions with other organisms within environmental communities. Regarding the ocean ecosystem, single-cell genomics has revealed high levels of genomic microdiversity within SAR11. While these genomic differences may seem subtle, they help subpopulations occupy distinct ecological niches, highlighting the importance of these genetic differences. The continuing development of approaches that link SCG to a combination of Raman spectroscopy and microfluidic cultivation devices has made it possible to correlate metabolic activity with genomic information and isolate cells of interest with specific functions, to derive an overall greater level of resolution in microbiological studies. The developing improvements to linked-read sequencing and amplification methods are, in many ways, driving down the cost of single-cell genomics and providing a format accessible to many scientists who are collecting extensive amounts of genomic data, and bringing us closer to understanding previously unimagined dimensions of microbial diversity and function across different ecological contexts (Dubey et al., 2019).

REFERENCES

Adeniji, A., Huang, J., Li, S., Lu, X., & Guo, R. (2024). Hot viewpoint on how soil texture, soil nutrient availability, and root exudates interact to shape microbial dynamics and plant health. *Plant and Soil*, 1–22.

Alatraktchi, F. A. A., Svendsen, W. E., & Molin, S. (2020). Electrochemical detection of pyocyanin as a biomarker for Pseudomonas aeruginosa: A focused review. *Sensors*, *20*(18), 5218.

Al-Hinai, M. A., Jones, S. W., & Papoutsakis, E. T. (2015). The Clostridium sporulation programs: Diversity and preservation of endospore differentiation. *Microbiology and Molecular Biology Reviews*, *79*(1), 19–37.

Amom, T., & Nongdam, P. (2017). The use of molecular marker methods in plants: A review. *International Journal of Current Research and Review*, *9*(17), 1–7.

Antil, S., Abraham, J. S., Sripoorna, S., Maurya, S., Dagar, J., Makhija, S., . . . Toteja, R. (2023). DNA barcoding: An effective tool for species identification—a review. *Molecular Biology Reports*, *50*(1), 761–775.

Bartram, A. K., Lynch, M. D., Stearns, J. C., Moreno-Hagelsieb, G., & Neufeld, J. D. (2011). Generation of multimillion-sequence 16S rRNA gene libraries from complex microbial communities by assembling paired-end Illumina reads. *Applied and Environmental Microbiology*, *77*(11), 3846–3852.

Beveridge, T. J. (2001). Use of the gram stain in microbiology. *Biotech Histochem*, *76*(3), 111–118.

Bouchez, T., Blieux, A. L., Dequiedt, S., Domaizon, I., Dufresne, A., Ferreira, S., . . . Ranjard, L. (2016). Molecular microbiology methods for environmental diagnosis. *Environmental Chemistry Letters*, *14*(4), 423–441.

Cheng, C., Sun, J., Zheng, F., Wu, K., & Rui, Y. (2014). Molecular identification of clinical "difficult-to-identify" microbes from sequencing 16S ribosomal DNA and internal transcribed spacer 2. *Annals of Clinical Microbiology and Antimicrobials*, *13*(1), 1.

Choudhary, P., Singh, B. N., Chakdar, H., & Saxena, A. K. (2021). DNA barcoding of phytopathogens for disease diagnostics and bio-surveillance. *World Journal of Microbiology and Biotechnology*, *37*(3), 54.

Church, D. L., Cerutti, L., Gürtler, A., Griener, T., Zelazny, A., & Emler, S. (2020). Performance and application of 16S rRNA gene cycle sequencing for routine identification of bacteria in the clinical microbiology laboratory. *Clinical Microbiology Reviews*, *33*(4), 10–1128.

Dubey, R. K., Tripathi, V., Prabha, R., Chaurasia, R., Singh, D. P., Rao, C. S., . . . Abhilash, P. C. (2019). Single-cell genomics and metagenomics for microbial diversity analysis. In *Unravelling the Soil Microbiome: Perspectives For Environmental Sustainability* (pp. 33–49). Cham: Springer International Publishing.

Earth, M. D. B. O. (2018). Ecological significance of microbial trophic mixing in the oligotrophic ocean: The Atlantic Ocean case studies. *Microbial Ecology of the Oceans*, 99.

Fakruddin, M. D., & Mannan, K. S. B. (2013). Methods for analyzing diversity of microbial communities in natural environments. *Ceylon Journal of Science (Biological Sciences)*, *42*(1).

Fjukstad, B., & Bongo, L. A. (2017). A review of scalable bioinformatics pipelines. *Data Science and Engineering*, *2*(3), 245–251.

Franco, C. M., Ambrosio, R. L., Cepeda, A., & Anastasio, A. (2021). Fish intended for human consumption: From DNA barcoding to a next-generation sequencing (NGS)-based approach. *Current Opinion in Food Science*, *42*, 86–92.

Fukuda, K., Ogawa, M., Taniguchi, H., & Saito, M. (2016). Molecular approaches to studying microbial communities: Targeting the 16S ribosomal RNA gene. *Journal of UOEH*, *38*(3), 223–232.

Hafez, E. E., & Elbestawy, E. (2009). Molecular characterization of soil microorganisms: Effect of industrial pollution on distribution and biodiversity. *World Journal of Microbiology and Biotechnology*, *25*(2), 215–224.

Harari, O., Park, S. Y., Huang, H., Groisman, E. A., & Zwir, I. (2010). Defining the plasticity of transcription factor binding sites by deconstructing DNA consensus sequences: The PhoP-binding sites among gamma/enterobacteria. *PLoS Computational Biology*, *6*(7), e1000862.

Hayat, R., Ali, S., Amara, U., Khalid, R., & Ahmed, I. (2010). Soil beneficial bacteria and their role in plant growth promotion: A review. *Annals of Microbiology*, *60*(4), 579–598.

Islam, W., Noman, A., Naveed, H., Huang, Z., & Chen, H. Y. (2020). Role of environmental factors in shaping soil microbiome. *Environmental Science and Pollution Research*, *27*(33), 41225–41247.

Joly, S., Davies, T. J., Archambault, A., Bruneau, A., Derry, A., Kembel, S. W., . . . Wheeler, T. A. (2014). Ecology in the age of DNA barcoding: The resource, the promise and the challenges ahead. *Molecular Ecology Resources*, *14*(2), 221–232.

Kalam, S., Basu, A., Ahmad, I., Sayyed, R. Z., El-Enshasy, H. A., Dailin, D. J., & Suriani, N. L. (2020). Recent understanding of soil acidobacteria and their ecological significance: A critical review. *Frontiers in Microbiology*, *11*, 580024.

Kawaka, F. (2022). Characterization of symbiotic and nitrogen fixing bacteria. *AMB Express*, *12*(1), 99.

Khatoon, H., Solanki, P., Narayan, M., Tewari, L., Rai, J. P. N., & Hina Khatoon, C. (2017). Role of microbes in organic carbon decomposition and maintenance of soil ecosystem. *International Journal of Chemical Studies*, *5*(6), 1648–1656.

Kim, D. H., Lee, B. K., Kim, Y. D., Rhee, S. K., & Kim, Y. C. (2010). Detection of representative enteropathogenic bacteria, Vibrio spp., pathogenic Escherichia coli, Salmonella spp., Shigella spp., and Yersinia enterocolitica, using a virulence factor gene-based oligonucleotide microarray. *The Journal of Microbiology*, *48*(5), 682–688.

Krawczyk, B., & Kur, J. (2018). Molecular identification and genotyping of staphylococci: Genus, species, strains, clones, lineages, and interspecies exchanges. In *Pet-to-Man Travelling Staphylococci* (pp. 199–223). Academic Press.

Li, C., Quan, Q., Gan, Y., Dong, J., Fang, J., Wang, L., & Liu, J. (2020). Effects of heavy metals on microbial communities in sediments and establishment of bioindicators based on microbial taxa and function for environmental monitoring and management. *Science of the Total Environment*, *749*, 141555.

Lund, P. A. (2009). Multiple chaperonins in bacteria–why so many? *FEMS Microbiology Reviews*, *33*(4), 785–800.

Maiden, M. C. (2006). Multilocus sequence typing of bacteria. *Annual Review of Microbiology*, *60*(1), 561–588.

Maurer, J. J. (2011). Rapid detection and limitations of molecular techniques. *Annual Review of Food Science and Technology*, *2*(1), 259–279.

Mishra, P., Kumar, A., Nagireddy, A., Mani, D. N., Shukla, A. K., Tiwari, R., & Sundaresan, V. (2016). DNA barcoding: An efficient tool to overcome authentication challenges in the herbal market. *Plant Biotechnology Journal*, *14*(1), 8–21.

Mohammadi, K., Heidari, G., Khalesro, S., & Sohrabi, Y. (2011). Soil management, microorganisms and organic matter interactions: A review. *African Journal of Biotechnology*, *10*(86), 19840.

Naqvi, S. A. H., Wang, J., Malik, M. T., Umar, U. U. D., Hasnain, A., Sohail, M. A., . . . Datta, R. (2022). Citrus canker—distribution, taxonomy, epidemiology, disease cycle, pathogen biology, detection, and management: A critical review and future research agenda. *Agronomy*, *12*(5), 1075.

Oehler, J. B., Wright, H., Stark, Z., Mallett, A. J., & Schmitz, U. (2023). The application of long-read sequencing in clinical settings. *Human* genomics, *17*(1), 73.

Pavoine, S., & Bonsall, M. B. (2011). Measuring biodiversity to explain community assembly: A unified approach. *Biological Reviews*, *86*(4), 792–812.

Poirier, S., Rué, O., Peguilhan, R., Coeuret, G., Zagorec, M., Champomier-Verges, M. C., . . . Chaillou, S. (2018). Deciphering intra-species bacterial diversity of meat and seafood spoilage microbiota using gyrB amplicon sequencing: A comparative analysis with 16S rDNA V3-V4 amplicon sequencing. *PLoS One*, *13*(9), e0204629.

Pontes, D. S., Lima-Bittencourt, C. I., Chartone-Souza, E., & Amaral Nascimento, A. M. (2007). Molecular approaches: Advantages and artifacts in assessing bacterial diversity. *Journal of Industrial Microbiology and Biotechnology*, *34*(7), 463–473.

Rajguru, B., Shri, M., & Bhatt, V. D. (2024). Exploring microbial diversity in the rhizosphere: A comprehensive review of metagenomic approaches and their applications. *3 Biotech*, *14*(10), 224.

Ramesh, M., Sen, A., Vachher, M., & Nigam, A. (2021). Delineating bacteria using DNA barcoding. *Molecular Genetics, Microbiology and Virology*, *36*(Suppl 1), S65–S73.

Ranawat, P., & Rawat, S. (2017). Stress response physiology of thermophiles. *Archives of Microbiology*, *199*(3), 391–414.

Rave, A. F. G., Kuss, A. V., Peil, G. H. S., Ladeira, S. R., Villarreal, J. P. V., & Nascente, P. S. (2018). Biochemical identification techniques and antibiotic susceptibility profile of lipolytic ambiental bacteria from effluents. *Brazilian Journal of Biology*, *79*(4), 555–565.

Ren, B., Hu, Y., Chen, B., Zhang, Y., Thiele, J., Shi, R., . . . Bu, R. (2018). Soil pH and plant diversity shape soil bacterial community structure in the active layer across the latitudinal gradients in continuous permafrost region of Northeastern China. *Scientific Reports*, *8*(1), 5619.

Sanger, F., Nicklen, S., & Coulson, A. R. (1977). DNA sequencing with chain-terminating inhibitors. *Proceedings of the National Academy of Sciences*, *74*(12), 5463–5467.

Sheth, B. P., & Thaker, V. S. (2017). DNA barcoding and traditional taxonomy: An integrated approach for biodiversity conservation. *Genome*, *60*(7), 618–628.

Sousa, A. M., Machado, I., Nicolau, A., & Pereira, M. O. (2013). Improvements on colony morphology identification towards bacterial profiling. *Journal of Microbiological Methods*, *95*(3), 327–335.

Stal, L. J. (2007). Cyanobacteria: Diversity and versatility, clues to life in extreme environments. In *Algae and Cyanobacteria in Extreme Environments* (pp. 659–680). Dordrecht: Springer Netherlands.

Stewart, E. J. (2012). Growing unculturable bacteria. *Journal of bacteriology*, *194*(16), 4151–4160.

Suvorov, M., Fisher, J. F., & Mobashery, S. (2008). Bacterial cell wall: Morphology and biochemistry. In *Practical Handbook of Microbiology* (pp. 179–210). CRC Press.

Thanassi, D. G., Bliska, J. B., & Christie, P. J. (2012). Surface organelles assembled by secretion systems of Gram-negative bacteria: Diversity in structure and function. *FEMS Microbiology Reviews*, *36*(6), 1046–1082.

Tilak, K. V. B. R., Ranganayaki, N., Pal, K. K., De, R., Saxena, A. K., Nautiyal, C. S., . . . Johri, B. N. (2005). Diversity of plant growth and soil health supporting bacteria. *Current Science*, 136–150.

Tshikhudo, P., Nnzeru, R., Ntushelo, K., & Mudau, F. (2013). Bacterial species identification getting easier. *African Journal of Biotechnology*, *12*(41).

Vandamme, P., Pot, B., Gillis, M., De Vos, P., Kersters, K., & Swings, J. (1996). Polyphasic taxonomy, a consensus approach to bacterial systematics. *Microbiological Reviews*, *60*(2), 407–438.

Varadi, L., Luo, J. L., Hibbs, D. E., Perry, J. D., Anderson, R. J., Orenga, S., & Groundwater, P. W. (2017). Methods for the detection and identification of pathogenic bacteria: Past, present, and future. *Chemical Society Reviews*, *46*(16), 4818–4832.

Voigt, B., Schweder, T., Becher, D., Ehrenreich, A., Gottschalk, G., Feesche, J., . . . Hecker, M. (2004). A proteomic view of cell physiology of Bacillus licheniformis. *Proteomics*, *4*(5), 1465–1490.

Volokhov, D. V., Simonyan, V., Davidson, M. K., & Chizhikov, V. E. (2012). RNA polymerase beta subunit (rpoB) gene and the 16S–23S rRNA intergenic transcribed spacer region (ITS) as complementary molecular markers in addition to the 16S rRNA gene for phylogenetic analysis and identification of the species of the family Mycoplasmataceae. *Molecular Phylogenetics and Evolution*, *62*(1), 515–528.

Wadt, D. (2024). *Development and Testing of Selective and Differential Media for the Recovery of Bacterial Bovine Respiratory Disease Isolates from Feedlot water Bowls* (Doctoral dissertation).

Wani, G. A., Khan, M. A., Dar, M. A., Shah, M. A., & Reshi, Z. A. (2021). Next generation high throughput sequencing to assess microbial communities: An application based on water quality. *Bulletin of Environmental Contamination and Toxicology*, *106*(5), 727–733.

6 DNA Barcoding and Marine Biodiversity

Muhammad Hassan Jahanzaib, Nazia Ehsan, and Hayssam Mohamed Ali

6.1 INTRODUCTION

The modern world must address several critical biological challenges, such as preserving biodiversity, assuring biosecurity, protecting species, and preventing pandemics. The attainment of these objectives, along with the efficacy of the ensuing plan of action, necessitates effective overall networks and depends on our ability to properly classify any specified species (Frézal and Leblois, 2008). As Dayrat stated, 'Accurately defining species borders and identifying species are essential for understanding life's diversity, as they determine whether distinct organisms fit into the same species or not.' Identification of species relies on taxonomists' experiences, whose work fails to cover all the taxon identifications necessary for unskilled users. The 'DNA Barcode of Life' initiative seeks to establish a consistent, quick, and economical approach for species recognition that is accessible to non-specialists (DeLong, 2004).

Identifying organisms based on physical characteristics typically poses a tough process that necessitates skilled taxonomists. Based on morphology, techniques generally are prolonged and may not consistently resolve at the species level. Molecular investigations have uncovered various living species that have undergone genetic variation and lack corresponding phenotypic differences, rendering them indistinguishable from the traditional morphological species concept. Identifying morphologically cryptic species presents a main task in recent taxonomy (Heinrichs *et al.*, 2011). Animals pass through different changing life cycles, which include many morphologically different phases. Sometimes, these species show morphological similarities with the full stage of other species. The specimens are often compromised and provide few tissue samples for identification, making morphological classification improbable.

It also applies to several plants whose vegetative phases (intervals devoid of blooms or fruits) hinder taxonomic clarity (Fišer Pečnikar and Buzan, 2014). In the 1990s, molecular identification also arose with the progress of the PCR technique. For bacterial identification, molecular identification was also used. Molecular identification has primarily remained used in bacterial studies, microbial biodiversity assessments, and the routine diagnosis of pathogenic strains (Wirth *et al.*, 2006; Zhou *et al.*, 1997).

DNA barcoding involves extracting small DNA sequences that aid in identifying and discovering species. The prime target of DNA barcoding is classifying

DOI: 10.1201/9781003538929-6

unidentified samples according to established taxonomy (Schindel and Miller, 2005). DNA barcoding is an innovative technique in taxonomy and is considered a valid method for arranging specimens and their data in systematic studies.

Organizing specimens by barcode haplotype from the outset of the study allows efficient material inspection and enhances the organization and preservation of extensive character information on data and life history, contingent on data availability for the barcoded specimens (DeSalle and Goldstein, 2019). The diversity of species that have been described up until this point constitutes a very small fraction of the total biodiversity on Earth (roughly 9 million) (Rico-Sánchez *et al.*, 2020). It is, therefore, extremely challenging to estimate accurately the variety of life forms that become extinct daily, given that scientists have described only 10–15% of the planet's total diversity (Rico-Sánchez *et al.*, 2020). Due to several issues, including an absence of finances for taxonomy, limited skilled taxonomists, and the unavailability of precise species identification methods, a significant amount of the diversity, approximately 86–91% (equivalent to approximately 7.2 million), is still not identified (Park *et al.*, 2019). The procedure entails the accumulation of samples from the field, the extraction of DNA, the choice of the barcoding gene to be amplified through the utilization of a universal primer, and the sequencing of the amplified DNA molecule through NGS and Sanger sequencing. The purpose of evaluating diversity and investigating data obtained using data analysis tools like Mother, Qiime2, and other similar programs (Sire *et al.*, 2019). DNA barcoding has become increasingly precise, faster, and authentic due to the introduction of contemporary sequencing techniques. The techniques include ONT, Pac Bio, MinION Nanopore sequencing, and NGS methods (Antil *et al.*, 2023). Oceans exhibit greater biodiversity than terrestrial and freshwater environments, as they account for more than 70% of the Earth's surface. Cryptic organisms are prevalent in marine habitats.

Cryptic species are morphologically analogous yet genetically divergent. DNA barcoding is valuable for estimating these cryptic species (Trivedi *et al.*, 2016a). DNA barcoding aims to establish an effective, correct, and representative reference database, which requires standardized and cost-efficient methods for DNA isolation, amplification, and sequencing.

In the long run, the reference library will enhance its utility, facilitating the swift identification of taxa at lower taxonomic levels using specific short-DNA sequences (Hajibabaei *et al.*, 2006). The oceanic environment plays a significant role in the history of life, and the present sea maintains its fundamental role in biological and climatic evolution (Boeuf, 2011). The presence of cryptic species in aquatic habitats underscores the significance of molecular techniques, such as DNA barcoding, for species identification (Molnar *et al.*, 2008). According to current estimations, 250,000 to 274,000 species have been identified.

Information technology has significantly facilitated the compilation and updating of species catalogs; nonetheless, we remain distant from achieving a comprehensive global inventory of organisms, especially within maritime habitats (Bouchet, 2006). Species identification and categorization have historically belonged to the field of taxonomists, establishing a classification system framework and serving as a fundamental for various biotic investigations. Contemporary society must address several critical biological challenges, including the preservation of biodiversity, assurance of biosecurity, the safety of species, and the prevention of pandemics. The attainment of these objectives and the efficacy of ensuing campaigns necessitate effective worldwide networks and depend on our ability to recognize any specified species (Frézal and Leblois, 2008). Life's existence and diversity are two of life's most remarkable features.

One of the most amazing things about the Earth is the presence of life. The word 'biological diversity' describes the variety found in living things from every type of ecosystem, including biological systems and terrestrial, marine, and other aquatic areas. Diversity within species, diversity between species, and diversity in ecosystems are included in this diversity (Planes *et al.*, 2020).

6.2 PRINCIPLES OF DNA BARCODING

The focus of molecular systematics in DNA barcoding is placed on single loci in the mitochondrial genome. This particular portion was chosen because it has homologous segments across taxa that can be used in primer design. Furthermore, it also contains variation at two levels: interspecific and intraspecific. This particular DNA sequence is designed to represent the 5′ end of gene encoding (COI) subunit 1. COI 1 is different enough to identify a large majority of animal species. Numerous research studies have proved that this method is efficient in the animal realm and has been applied to sample many types of tissues.

6.2.1 Genetic Markers Used in DNA Barcoding

A team from the University of Guelph in Canada, under the supervision of Prof. Paul Herbert, gave the idea of DNA barcoding in 2003. The genome region suggested for the barcode of mammals is the COI gene. (Trivedi *et al.*, 2016a). In the first international conference on DNA Barcoding, the sequence of this COI region was verified. Many studies have revealed that the sequence variability within a stretch of approximately 650 base pairs, close to the start of the COI gene, delivers extraordinarily resolved nodes for various animal groups, including birds (Schindel *et al.*, 2011). Other advantages can be gained from DNA barcoding, such as seafood safety evaluation and authentication, animal species identification in forensics, invasive alien species (IAS) identification, and conservation genetics.

Only a few global DNA barcoding programs operate in marine ecosystems, including MarBOL, CeDAMar, CMarZ, and SHARK-BOL. The mitochondrial gene (COl) sequence is recognized as an all-purpose barcode for identification (Trivedi *et al.*, 2016a). Animal species are inadequate for differentiating between the majority of plant species. This limitation arises from plant genomes' significantly slower mutation rate (Kress *et al.*, 2005). Given the substantial study devoted to investigating a universal plant barcode, it is noteworthy that no current loci have demonstrated efficacy across all plant species (Chen *et al.*, 2010).

The combination of two loci, rbcL and matK, is confirmed by CBOL as the most efficient plant barcode, with a differentiating efficiency of 72%. Taxonomists have suggested that employing a multi-locus approach could be crucial for accurately differentiating plant taxa (Li *et al.*, 2015).

The seemingly simple task of identifying an appropriate locus for application as a plant barcode has revealed itself to be considerably more intricate than initially expected, leading to considerable contention. At present, an agreement on a uniform barcode for plants remains elusive; however, ecologists, evolutionary biologists, taxonomists, and conservationists are engaged in thoughtful deliberations regarding the potential implementation of a genetic identifier in a variety of studies and applied initiatives (Lahaye *et al.*, 2008).

DNA barcodes must fit the gene region into three specific criteria:

1. It should be variable and diverse enough at the species level.
2. Conserved region identification is very important for designing the primers used for PCR. These primers can be applied across different taxa.
3. Maintaining a short sequence length is crucial for enhancing modern DNA extraction and amplification methods. The 600-base pair DNA sequence located within the mitochondrial gene encoding the mitochondrial gene CO1 remained identified as a credible and standardized barcode for species identification at the animal level (Kress and Erickson, 2008).

6.3 PROCEDURE FOR DNA BARCODING

DNA barcoding involves two key phases: The initial phase consists of creating a comprehensive barcode library with identified species. It is followed by comparing the barcode sequence from an unidentified model with the developed library to help identify it. The initial phase needs a deep understanding of ecological principles to effectively select one or more individuals from each species as reference models for the barcode library. Samples of tissues for DNA isolation can be taken from both living and preserved specimens in museums.

The samples are subjected to a series of laboratory procedures that encompass tissue sampling, DNA processing, and sequencing, ultimately resulting in the application of a DNA barcode, which is represented as a chromatogram. A chromatogram is a graphical illustration of a DNA sequence formed by a sequencing instrument. The barcode may be stored within a database for prospective utilization, or utilized as a query sequence for comparative analysis with sequences presently housed in the database (Kaur, 2015).

6.3.1 Sampling and Vouchering

The process of DNA barcoding represents a critically important phase in the overall methodology. The habitat serves as the origin of the biological samples that are collected. The potential for barcoding to be applied to museum samples also exists. The proper vouchering of samples is an essential aspect that warrants careful consideration. It is vital that the documentation provided is thorough and accompanies the samples that have been collected. Necessary details include the name of the collector, the date of collection, the geographical site (with specific latitude and longitude), the height and depth measurements, observations regarding the habitat, the voucher number, the specimen name which is identified, photographs, and any additional identifying information that may be pertinent. Different groups and categories of samples possess distinct requirements and data necessities, encompassing the need for vouchers.

6.3.2 Tissue Extraction

In the subsequent phase of the barcoding procedure, a minimal quantity of tissue that meets the necessary criteria is typically employed.

6.3.3 Extraction of DNA

The phenol–chloroform–isoamyl alcohol method is used as the appropriate procedure for DNA extraction. Subsequently, the extracted DNA is quantified using a spectrophotometric method, while agarose gel electrophoresis is employed to visualize the DNA.

6.3.4 Barcode Region Amplification

In the context of polymerase chain reaction (PCR) amplification aimed at the barcode sequence, researchers may employ either custom-designed primer sets or those that have been previously published in the literature. After that, the amplified products through PCR are then analyzed by agarose gel electrophoresis.

6.3.5 PCR Amplicon Purification and DNA Sequencing

The first process to remove the PCR-amplified products includes extracting them from the gel and sequencing this DNA.

6.3.6 How to Submit Sequencing Data to Global Repositories: GenBank and BOLD

Once the DNA sequence is successfully integrated, they are subsequently uploaded to an international database like (NCBI). The center will then assign the required accession numbers as needed.

6.3.7 Bioinformatics Analysis and Interpretation of Sequences

Sequence analysis is performed using various bioinformatics tools for species identification, molecular phylogenetic research, and the discovery of new species. Bioinformatics tools encompass a variety of applications, including BLAST, CLUSTAL, and others (Trivedi *et al.*, 2014).

6.3.8 Assessing the Precision of Species Identification

The main purpose of evaluating species identification precision is to assess the effectiveness of the associated barcode database. The barcode database operates through a set workflow. The DNA barcode obtained from the specimen and the sample's barcode sequence are then used to query the database. The database's software subsequently provides a potential species identification. Theoretically, the absence of a species within the database could lead to its classification as an 'unknown species.'

There is currently a lack of agreement within the scientific community concerning the specific criteria that should be employed to describe species borders through DNA barcoding techniques. As a result, the database output indicating 'unknown species' (that is, 'beyond the limits of every recognized species boundary') poses significant challenges. Consequently, to enhance clarity in the instructional process, this chapter is dedicated solely to identifying recognized species, assuming each sample referenced in the barcode database aligns with a species already documented within that database (Trivedi *et al.*, 2014).

Species boundaries are frequently linked to variations in DNA, exemplified by a 2% difference (i.e., a p-distance of 0.98); nonetheless, specific limit distinctions may differ across various species. The phenomenon of incomplete lineage sorting has the potential to obliterate the distinction. Many endeavors to reinterpret species' boundaries by examining threshold differences (Lambert *et al.*, 2005) substantially opposed the established taxonomy (Meier *et al.*, 2006).

6.3.9 Using DNA Barcoding for Marine Biodiversity Research

DNA barcoding offers considerable advantages in species identification and classification at the species level, especially where traditional morphological methods are puzzling or when taxa exhibit morphological ambiguity.

This is relevant to cryptic species found inside plant and animal kingdoms. In recent years, many publications addressing this topic have surfaced; however, we shall refrain from conducting an in-depth examination of these works, and instead provide a curated selection of notable examples. An interesting example is identifying 810 samples associated with long-eared bats (3 species) living in Western Europe (Andriollo and Ruedi, 2020).

DNA barcode markers allow distinguishing between narrowly related species and cultivars in certain conditions (Punina *et al.*, 2017).

Molecular analysis employing DNA barcodes has been utilized to distinguish between different varieties of *Tabernaemontana divaricata* (L.) R. Br., a vibrant green shrub. They assessed seven genetic markers: tmH-psbA, rpoC, matK, rpoB, and psbK-psbI. Five loci, rpoB, rbcL, psbK, atpF-atpH, and trnH-psbA, were successfully obtained and used for further analysis. More importantly, out of these five, only two markers provided enough taxonomic distinction to differentiate the cultivars of *T. divaricate* (Jena *et al.*, 2019).

This study possesses significant relevance and is representative of the marker search effort. As in many articles devoted to DNA barcoding, the marker search details can often be presented insufficiently in many studies concerning DNA barcoding. This oversight leads to the inclusion of only sequences successfully utilized as DNA barcodes, thereby restricting access to a wealth of pertinent information (Shadrin, 2021).

6.3.10 Marine Species Identification

First, it is crucial to create a thorough database that includes specimens and their respective marker sequences, and species binomials. Using species binomials serves as a definitive benchmark for species identification, necessitating a high degree of accuracy to prevent the misclassification of correctly identified species (Trivedi *et al.*, 2014).

Next, choosing an appropriate computational algorithm is essential for linking species to their barcode sequences. BLAST + P demonstrated notable improvement over the basic method of classifying specimens based on species of databank sequence(s) with the closest p-distance. Global sequence alignments, performed using algorithms like Needleman–Wunsch, or several other sequence alignment methods, enable the alignment of entire barcode sequences by utilizing all available data.

This comprehensive approach often results in more precise species identifications compared to local sequence alignments, exemplified by BLAST, particularly in instances where BLAST generates arrangements of limited subsequences inside the larger sequences (Trivedi *et al.*, 2014). On the designation of the (PCI) 'The Probability of correct identification' consensus arises (Trivedi *et al.*, 2014).

6.4 MARINE FUNGI DNA BARCODING

Fungi play an important and diverse part in the biosphere, performing numerous essential functions across dissimilar environments. Obligate marine fungi exclusively live and reproduce in oceanic or estuarine environments. Facultative marine fungi, on the other hand, originate from freshwater or terrestrial habitats but can adapt to survive and reproduce in marine environments.

Marine fungi fulfill a crucial ecological function as decomposers and pathogens, and establish symbiotic relationships with aquatic organisms. It has been estimated that about 1500 marine fungi species exist in the world's marine environment, excluding fungi that act symbiotically with lichen. Therefore, DNA barcoding offers a useful approach for fast and precise worldwide assessment of marine fungal species. (Velmurugan *et al.*, 2013; Evans-Illidge *et al.*, 2015). Fungi identification through appearance often entails cultivating the specific fungus and examining its properties.

This procedure can be very time-consuming and often involves some form of hurdle (O'Brien *et al.*, 2005). Evaluation of shape, size, and other features of the spore plays one of the central roles in species identification, which requires a professional level of knowledge. Therefore, molecular approaches like DNA barcoding could be only operative in recognizing such creatures (Hebert *et al.*, 2003).

6.5 MARINE ANIMAL DNA BARCODING

Marine animals come in all forms, astounding and ranging within an extreme size class: microorganisms at one pole and gigantic animals at the other. There are 35 recognized animal phyla; all but one have marine representatives, and 14 phyla are exclusively marine (Briggs, 1994).

6.5.1 Marine Zooplanktons

Zooplankton are of significant ecological importance and are composed of 15 animal phyla. As a result, DNA barcoding for zooplankton has become an essential component of up-to-date ecological research. The Census of Marine Zooplankton (CMarZ) seeks to offer a wide-ranging study of zooplankton communities across different parts of oceans worldwide. Copepods belong to the phylum Arthropoda and are among the most diverse and widely studied zooplankton in aquatic environments.

Several barcoding analyses of copepod species from marine environments have been conducted, and these have presented many new species and improved knowledge of the undesirable species complexity and the distribution of various copepod groups in different geographical areas (Blanco-Bercial *et al.*, 2011).

The average K2P distance seen in the species was calculated to be 0.0145. Copepods are the most diverse and relevant subgroup of marine zooplankton, including over 2500 species. This diverse group has been extensively studied in numerous research projects. Cryptic species in the copepod population require additional study using DNA barcoding techniques (Trivedi *et al.*, 2016a).

Amphipod zooplanktons, which belong to the phylum Arthropoda, have also been studied. The four molecular barcodes used to distinguish *Neocalanus* copepods were 28S, nuclear ITS, 12S, and COI (Machida and Tsuda, 2010). Barcoding of holozooplankton specimens from the Arctic Ocean and metazoan zooplankton samples from the Sargasso Sea was performed as detailed in numerous publications outlined here (Bucklin *et al.*, 2010). In this work, the chosen method—DNA barcoding—has been proven to be particularly effective in recognizing distinct species of chaetognaths, which are quite tough for morphological analysis (Jennings *et al.*, 2010). The approach to analyzing barcodes through the COI gene in 52 specimens, including fourteen recognized species of chaetognaths, successfully differentiated between the species in the phylum Chaetognatha.

6.5.2 Marine Invertebrates

Using DNA barcoding for various classes of oceanic invertebrates has enhanced our understanding of these species and their distribution in marine environments.

6.5.2.1 Porifera

Also, sponges are important in the Antarctic Ocean, functioning as the substrate for most benthic species. A few Antarctic sponges were barcoded recently, and the information obtained helped researchers learn more about the vast variety of sponges that dwell in that part of the world (Vargas *et al.*, 2015). The barcoding approach was recently used on lithiated demosponges, providing valuable information regarding their evolution. The authors have proposed reallocating various groups within this classification (Schuster *et al.*, 2015). The comprehensive analysis of the whole mitochondrial genome of *Aplysina cauliform* has unveiled specific regions that warrant further investigation in conjunction with COI sequences, particularly in the context of DNA barcoding for this verongida sponge (Sperling *et al.*, 2012).

6.5.2.2 Cnidaria

Monitoring the distribution and abundance of jellyfish is crucial for identifying non-native species that negatively impact the local community. Due to global warming, the number of jellyfish has increased exponentially, causing economic damage globally. Research was done in South Korea on marine jellyfish, in which microarray analysis and DNA barcoding were combined to create a DNA chip used for jellyfish recognition (Lee *et al.*, 2011). The ideal coral species, *Stylophora pistillate*, has shown different clades and notable changes in this coral 'lab-rat' (Keshavmurthy *et al.*, 2013).

6.5.2.3 Annelida

Similar to polychaetes, Annelida includes segmented marine worms. In the genus *Grania*, belonging to Annelida, four new species were discovered with the help of a study directed in the Great Barrier Reef (De Wit *et al.*, 2009). The competence of DNA barcoding is confirmed in several studies. Marine polychaete species majorly affect the identification (Lobo *et al.*, 2016).

6.5.2.4 Platyhelminthes

Some barcoding studies have been done on flatworms and marine flatworms. The problem with flatworm barcoding is that universal primers are unavailable for diversified flatworms (Vanhove *et al.*, 2013).

6.5.2.5 Nemathelminths

Nematodes play a crucial role as marine pollution indicators. The 18S gene is used with remarkable efficacy for barcoding marine nematodes (Bhadury *et al.*, 2006). COI, ITS, and D2D3 sequences are used for molecular phylogeny studies.

6.5.2.6 Mollusca

Aquatic mollusk DNA barcoding from Canada shows many unnoticed species. DNA barcoding of aquatic mollusks from Canada shows many unnoticed species (Layton *et al.*, 2014). DNA barcoding of 227 Canadian marine mollusk species revealed potential instances of unrecognized species (Layton *et al.*, 2014). Marine oysters, known for their edibility, are cherished as a gourmet seafood delicacy. Shell morphology of oysters is used for their identification. However, shell structure changes with habitat changes, making recognizing species difficult (Tack *et al.*, 1992). DNA barcoding resolves this problem very efficiently. It aids in the identification of marine bivalves and also benefits the understanding of their phylogenetic relationships (Trivedi *et al.*, 2015).

Marine oysters are bivalves of considerable commercial importance. Oyster recognition mainly depends on phenotypic traits like shell morphology, a process complicated by taxonomic disagreements. The habitat heavily influences shell morphology, seen as the primary distinguishing feature. Molecular identification is advantageous (Ward *et al.*, 2008). In the phylum Mollusca, in the class Gastropoda, pteropod species are famous in research because of their susceptibility to marine acidification. The *Diacovolinia* pteropods in Atlantic specimens are species under the same monophyletic clade. Differences in Pacific and Atlantic populations may exist at the species level (Maas *et al.*, 2013).

6.5.2.7 Arthropoda

The subphylum Crustacea encompasses a remarkably diverse array of marine species, while the class Pycnogonida is characterized by its exclusive representation of marine sea spiders. Merostomata, entirely marine, includes the living fossil horseshoe crab. These COI sequences of mitochondria were used to determine the phylogenetic tree of the horseshoe crabs. (Kamaruzzaman *et al.*, 2011).

6.5.2.8 Echinodermata

Echinoderms, therefore, are an unusual class of solely marine invertebrates that are conveniently distinguished by radial symmetry. This diverse phylum is separated into five living classes: Ophiuroidea, Crinoidea, Holothuroidea, Echinoidea, and Asteroidea. An extensive molecular analysis represented by five classes of echinoderms revealed that DNA barcoding was very successful in species identification (Ward *et al.*, 2008).

All echinoderms are marine animals; hence, they can only be found in seawater. DR analysis was conducted on 191 species of echinoderms, which were divided into five groups during the investigation on DNA barcoding. According to shallow within-species and deep between-species differences, 97.9% of these specimens remained correctly assigned to known species (Ward *et al.*, 2008). Similarly, an additional DNA barcoding for several different commercially important echinoderms was also performed (Uthicke *et al.*, 2010).

6.6 LOWER CHORDATES

All aquatic organisms include lower chordates, such as cephalochordates, hemichordates, and urochordates. Ascidians, which fit in the phylum Urochordata, are commonly considered benchmark model organisms for studying difficult biological mechanisms. These include an immune response, regulation of gene expression, heavy metal uptake and toxicity, cardiac and embryonic development, and differentiation. Ascidians are oceanic filter-feeding urochordates widely viewed as model organisms for studying difficult biological phenomena. They are used to examine transcriptional regulatory processes during embryonic growth, metal accumulation mechanisms, the evolutionary history of immunity, the preservation of gene regulatory networks in chordates, and aspects of cardiac development, and other subjects of interest (Razy-Krajka *et al.*, 2014).

Ciona intestinalis, the smallest chordate suitable for experimental manipulation, has been the focus of numerous genome analysis studies on its genetic material. These solitary ascidians were first identified through COI gene analysis of *Ciona* samples collected in New Zealand. *Savignyi Ciona*, found in the Southern Hemisphere, was first recorded, and molecular identification of *Ciona savignyi* in the Southern Hemisphere used the DNA barcoding technique (Smith *et al.*, 2012). A new species of ascidian, originally unreported, fitting to the family *Diplosomatidae* in the

Ryukyu Archipelago in Japan, has been identified using DNA barcoding methods (Hirose and Hirose, 2009).

6.7 VERTEBRATES

6.7.1 Fishes

Marine fish, rich in proteins, vitamin D, vitamin B12, iodine, selenium, and omega-3 fatty acids, form a crucial part of the human diet. The marine fisheries are of prime significance for food production and improving economic status. Accurate classification ensures effective fisheries management and authenticates food products for consumers. DNA barcoding can be understood as an easy, quick, and real means of identifying fish species. A wide and in-depth program for barcode projects for all the fish categories was begun (Ward *et al.*, 2009).

Two major worldwide efforts exist to barcode fish, including FISH-BOL (Trivedi *et al.*, 2016a). A comprehensive initiative has been initiated to implement a barcode system for all fish species (Ward *et al.*, 2009). Two major international initiatives are dedicated to the barcoding of fish species: FISH-BOL and SharkBOL. Many barcoding approaches examining fish species have been directed at diverse oceans and seas, including the Atlantic Ocean (Weigt *et al.*, 2012). The Arctic Ocean, the Pacific Ocean (Zemlak *et al.*, 2009), and the Indian Ocean (Dhaneesh *et al.*, 2015)

Thus, DNA barcoding is extremely effective at identifying diverse life forms and sizes of fish, not just the mature organisms but also the spawn, fresh fillets, fins, uniquely challenged specimens, and even broken pieces of a fish's body. This molecular method was applied to identify the source of shark fins confiscated from illegitimate fishing practices in Australia (Holmes *et al.*, 2009).

Studies on fish species within the genus *Sebastes* obtained from the China Sea provided evidence of the viability of using DNA barcoding to resolve species identities (Zhang *et al.*, 2013). The global demand for ornamental fish is quickly growing. A detailed study of the COI gene in 391 ornamental fish species from eight coral reef sites found that 98% were grouped into different barcode clusters.

6.7.2 Amphibians

Limited research has been undertaken regarding the use of barcoding techniques on amphibians. A recent initiative called the 'Cold Code' has started to apply barcoding to cold-blooded vertebrates, particularly amphibians and reptiles. Regarding the topic, it has been published that the 16S rRNA gene can act as the universal barcode for amphibians (Vences *et al.*, 2005).

6.7.3 Reptiles

Reptiles, which are cold-blooded vertebrates, are classified into four distinct orders: Crocodilians, Lizards, Turtles, and Tuatara. Understudying marine testudinal has shown that DNA barcoding suits species identification, phylogenetic assessment, and applying pertinent laws protecting endangered turtles (Naro-Maciel *et al.*, 2010).

A devoted study was directed to develop DNA barcoding methods assessing globally endangered marine turtles.

The current work shows that DNA barcoding efficiently identifies species and can significantly contribute to animal forensic and conservation genetics developments (Naro-Maciel *et al.*, 2010). A study done on Brazilian sea turtles shows that the program using the COI barcode tags has the potential to identify all the marine turtle species studied (Vargas *et al.*, 2009). While the barcoding effort involved 468 specimens fitting to the Squamata and Testudines orders in Madagascar, a hotspot country, the study has revealed new species (Nagy *et al.*, 2012).

Although the literature on fish DNA barcoding has been extensive, much less is known or researched about reptiles. The initial whole-genome DNA barcoding of reptiles, involving the nested orders Squamata and Testudines, was accomplished with 468 specimens collected from a biological diversity richness area, Madagascar. This study confirms the identification of 41 to 48 previously unknown species, demonstrating the efficiency of the DNA barcoding technique in evaluating species richness. The research concluded that the genetic distances across the families were primarily 13.4% for the Boidae and 29.8% for the Gekkonidae (Nagy *et al.*, 2012). A recently published paper is a good source for an overview of DNA barcoding in the dissimilar category class Reptilia (Trivedi *et al.*, 2016b).

6.7.4 Sea Birds

ABBI was developed in September 2005 to create a worldwide database of 10,000 recognized avian species. Molecular studies using DNA barcoding have established that several avian species exist that have not been documented scientifically.

ABBI was developed in September 2005 to create a worldwide database of 10,000 recognized avian species. Molecular studies using DNA barcoding have established that several avian species exist that have not been documented scientifically (Aliabadian *et al.*, 2013).

However, they live in the ocean or depend directly or indirectly on it, and still, seabirds cannot be considered marine inhabitants because they cannot use dissolved oxygen, which is present in water. There are Gulls, Terns, Brown Boobies, Carp Plovers, Swans, Spoonbills, and all sorts of the Ardeidae family and other kinds of seabirds. For food, the seabird counts on the ocean environment. Further, these species prefer nesting sites in sandy beaches, tidal flats, and mangrove habitats. More about this is explained by the fact that they build their nests in the mentioned habitats and mainly feed on fishes, shrimps, mollusks, and insects. Seabirds cannot survive when these essential habitats are not available. In the present study, eight genomic regions of Poland were targeted sequencing for DNA barcoding that included 387 specimens that belonged to 147 distinct avian species, such as seabirds commonly found in the Netherlands region. This country is considered one of the most significant on an international level if it speaks about numerous advancements made in the field of ornithology (Aliabadian *et al.*, 2013).

In certain regions of the world, avian species seeking refuge in mangroves are protected from predators. Land clearance for agricultural and other uses has also implicated mangroves in drastically reducing the habitats of some seabird species.

Concerning the issue of mangroves and seabirds, more related actions must be implemented in barcoding. Mangrove destruction, combined with the rise in oceanic acidity, salinity, and sea level, indicates the need for improved eccentric lobbying to conserve seabirds and deploy DNA bar codes (Trivedi *et al.*, 2014).

6.7.5 Mammals

A large barcoding survey carried out along the Atlantic coast of France has revealed the practicality of the presented method for tracking a variety of marine mammals (Alfonsi *et al.*, 2013).

6.8 OCEANIC PLANTS DNA BARCODING

Unlike observation on the animal kingdom, there is no clear concordance on which barcode gene is in the plant domain. Several molecular markers have been used to 'barcode' different groups of plants. Several researchers have devised a term known as the 'super barcode' to increase the efficiency of plant identification. The Plant Working Group of CBOL identifies and sets up standard barcode regions among plants. Indeed, the DNA barcoding system, which assesses the richness of plant species, has been subjected to methodological scrutiny (Trivedi *et al.*, 2016c).

6.8.1 Seagrasses

Due to their ability to suspend nutrients and their high primary production of seagrasses, other large floating plants most certainly have an impressive impact on the regional marine environment.

The most formidable problem in identifying seagrass species is the lack of flowers, which are considered valuable reference points for identification. Primers for DNA barcoding on the seagrasses were performed using several molecular markers, as outlined by ITS, trnK, rbcL, 5.8S rDNA, and matK (Lucas *et al.*, 2012).

While these seagrasses have been mapped and reported to have wide distribution across the global marine geographical map, the global map of seagrasses has revealed that their diversification and coverage have equally reduced. According to research, seagrass habitat has been shrinking to 110 square kilometers yearly since 1980. The substances contain essential secondary metabolites, such as phenolic acids, which are used in conventional medical conditions.

The antioxidant characteristics of rosmarinic acid and the antifouling attributes of zosteric acid—both sourced from seagrasses—are extensively employed across various industrial applications (Waycott *et al.*, 2009). Consequently, it is imperative to evaluate and preserve seagrasses immediately. Seagrasses exhibit both sexual and asexual means of reproducing, but the latter is more dominant; the former, however, are characterized by relatively low life spans.

Species recognition poses problems primarily due to the relatively restricted availability of flowers, which comprise a discrete morphological feature. The DNA barcoding methodology in this framework is highly efficient regarding identification questions. Various markers have been employed in identifying seagrasses, nuclear ITS, particularly for *Halophila* (Waycott *et al.*, 2002).

6.8.2 Marine Algae

Of all the genera, red algae from the *Gracilaria* genus were successfully DNA-barcoded (Kim *et al.*, 2010). DNA barcoding is an actual technique for quickly pinpointing HAB species and can be used for constant coastal and marine pollution assessment (Trivedi *et al.*, 2014). Marine algae are primarily categorized into two main groups: microalgae, which include cyanobacteria, dinoflagellates, and diatoms, and macroalgae, which are divided into green, brown, and red seaweeds. Microalgae, therefore, help as a valued source of various bioactive compounds. Seaweeds have also been used as nutritive value and are recognized to give out different gums. Applying the COI and UPA domain V of the 23S rRNA as molecular markers, Clarkston and Saunders disclosed a new species, namely *Euthoratim burtonii*. According to the outcomes of the present study, more optimal results were achieved through employing COI for species differentiation with the Kallymeniaceae family of red algae. An evaluation was therefore made to reflect on experiences learned and establish a conclusion that ascertained whether COI, UPA, and ITS facilitated the intertidal red macroalgae study (Zhao *et al.*, 2013).

6.8.3 Marine Phytoplankton

Phytoplankton plays a vital part in the complexity of coastal and oceanic environments. They inhabit the base trophic level of the oceanic food pyramid, serving as a vital link in transferring carbon dioxide from the atmosphere to the marine. Studying the distribution and availability of oceanic phytoplankton is a significant feature of ecological research. The photosynthetic rbcL gene was used as the molecular tag for the phytoplankton size classes below 10 μm (Bhattacharjee *et al.*, 2013).

6.8.4 Mangroves

Changes in salinity, nutrients, and dissolved oxygen value are occurring with time, directly affecting the distribution and density of different mangrove species in this region, which is already under pressure from human interference (Mitra *et al.*, 2015). Mangroves are a category of salt-tolerant plants that grow in the coastal areas of the tropical and subtropical zones. Coastal wetlands alone are known to have important ecological and economic impacts. They also play the important role of buffering against coastline areas most susceptible to intensive wave assault, storm flooding, and cyclonic motifs. Mangroves' ability to reduce the strength of coastal soil erosion is mainly due to the durable, steady root system of mangroves (Trivedi *et al.*, 2014). The nutrient, salinity level, and dissolved oxygen concentration show noticeable variation over time, resulting in various species of mangrove in this area being affected by human activities (Mitra *et al.*, 2015). DNA barcoding offers a unique approach that could help resolve issues related to these ecosystems' mangroves and associated flora and fauna (Trivedi *et al.*, 2014).

6.9 USE OF BARCODING TO DISTINGUISH BETWEEN CRYPTIC SPECIES AND CLOSELY RELATED SPECIES

Thus, an enormous diversity of cryptic species is available in aquatic or, specifically, marine habitats. It is proposed that DNA barcoding analyses can greatly improve the identification and assessment of hidden species (Pauls *et al.*, 2010). One of the researchers who contributed to the field did a study to show the significance of DNA barcoding, especially in recognizing cryptic species within the Hastings seabirds (Efe *et al.*, 2009).

Applying IAS also poses one of the main challenges for maintaining the richness and quality of the marine ecosystem. They are not only competitors with native species but are well-known economic challenges that should be resolved. Fouling is one of the activities that bring invasive species, and other practices like the discharge of ballast waters also bring invasive species that have become a major challenge to the Great Barrier Reef. The Caribbean tubeworm, known by the scientific nomenclature *Hydroides sanctaecrucis*, is an ideal example of such a tubeworm. One of the most effective techniques present in the approaches today is DNA barcoding, a method of quick recognition and assessment of invasive alien species (Molnar *et al.*, 2008)

6.9.1 Role of DNA Barcoding in Conducting Large-Scale Biodiversity Surveys

MarBOL, an abbreviation of Marine Barcode of Life, is a worldwide initiative dedicated to systematically barcoding a broad range of marine species from various taxa. It is a project linked to the Census of Marine Life (CoML). The CoML organization undertakes a diverse array of Ocean Realm Field Projects, encompassing the following initiatives, but not exclusively limited to.

The Tagging of Pacific Pelagics (TOPP), International Census of Marine Microbes (ICOMM), Natural Geography in Shore Areas (NaGISA), Pacific Ocean Shelf Tracking (POST), Global Census of Marine Life on Seamounts (CenSeam), Gulf of Maine Area Program (GOMA), Global Census of Coral Reef Ecosystems (CREEFS), Continental Margin Ecosystems on a Worldwide Scale (CoMargE), Census of Diversity of Abyssal Marine Life (CeDAMar), Census of Marine Zooplankton (CMarZ), Census of Antarctic Marine Life (CAML), Biogeography of Chemosynthetic Ecosystems (ChEss), Arctic Ocean Diversity (ArcOD), along with numerous additional projects, collectively contribute to the extensive understanding of marine biodiversity and ecosystem dynamics across various oceanic regions (Trivedi *et al.*, 2014).

Halophytic plants, also known as mangroves, are predominantly found in the seaside zones of hot and subtropical regions. Mangroves are significant in both economic and ecological contexts. They fulfill several essential functions, among which one of the most important is protecting fragile coastal regions from the damaging impacts of intense wave action, storm surges, and hurricanes. Mangroves possess a highly developed root system, which is central to coastal soil erosion (Trivedi *et al.*, 2014).

Salinity, nutrient levels, and dissolved oxygen concentrations are experiencing notable alterations over time, directly influencing the spreading and abundance of mangrove species in this area subjected to human-induced stressors (Mitra *et al.*, 2015). DNA barcoding can yield major insights into the assessment of mangrove biodiversity. Furthermore, it can also elucidate the relationships between mangroves and the various plant and wildlife species associated with these vital ecosystems. The implication of DNA barcoding for maintaining mangrove plants has been emphasized as a matter of considerable significance (Daru *et al.*, 2013). Sponges, belonging to the Phylum Porifera, are a varied group of sessile, benthic metazoans found in diverse environments, including marine, freshwater, and semi-terrestrial habitats.

Sponges fulfill crucial ecological functions in oceanic ecosystems, ranging from vibrant coral reefs to profound abyssal plains. There are 8,500 distinct species distributed across various aquatic ecosystems worldwide, encompassing transient freshwater environments, coastal regions, and the profound depths of the oceans. From a taxonomic and systematic perspective, the Phylum Porifera presents challenges due to the limited number of distinguishing features available for taxonomic and phylogenetic analysis of sponges (Erpenbeck *et al.*, 2006).

6.9.2 Conservation Efforts

The application of DNA barcoding greatly enhances the identification, evaluation, and preservation of marine biological resources, providing several highly significant advantages. The probabilistic identification of new species through DNA barcoding is considered one of the most significant accomplishments in biological science. Despite the relatively young age of the island of Madagascar, many new reptilian species have been discovered with the help of DNA barcodes there (Nagy *et al.*, 2012). Literature abounds with documents outlining the detection of new species using the DNA barcoding approach (De Wit *et al.*, 2009).

The genetic makeup of the endangered or threatened marine species can be practically determined using barcoding technology to identify the species. Besides helping properly identify these species, this technique goes a long way in waging the battle against the associated illegal trade, which includes body parts and larvae. With the help of DNA barcoding, it was easy to recognize shark fins sourced from poaching in the waters of Australia (Holmes *et al.*, 2009). In the same way, using DNA barcoding approaches with turtle flesh, carcasses, and eggs, the status of endangered sea turtles that were smuggled was investigated (Vargas *et al.*, 2009).

6.10 CHALLENGES AND LIMITATIONS

Yet this DNA barcoding tool does not come without a set of limitations, even though the method indeed resolves many issues of the traditional identification methods. A lack of a gene and universal primer in all living organisms, plus the large sequence variation that allows species' discrimination, is an important hindrance to the barcoding strategy and a weakness (Antil *et al.*, 2023)

There are few DNA barcode libraries with which comparisons are enabled. Further, a certain amount of quantitative data is lost because of the detrimental effects of the bias of primers and polymerase (Ji *et al.*, 2020). DNA barcoding allows species identification by examining genetic diversity within and between species.

Nevertheless, the boundaries of this variation remain inadequately comprehended and may differ across various taxonomic groups (Purty and Chatterjee, 2016).

The existence of heteroplasmy and pseudogenes significantly enhances databases' complexity while diminishing the accuracy of DNA barcoding technology. A pseudogene may cause an individual species to be misclassified into several distinct species. Pseudogenes can induce heteroplasmy, defined by the coexistence of various mitochondrial DNA (mtDNA) forms within a single organism. This phenomenon presents significant challenges in accurately identifying species with DNA barcoding techniques (Gong *et al.*, 2018).

The initial overview of DNA barcoding was met with resistance from conventional taxonomists (Rubinoff *et al.*, 2006). Several limitations are inherent to DNA barcoding methodology. These include challenges related to low resolution, particularly for species that have undergone recent divergence, as well as issues arising from species complexes and hybridization events. In their scholarly investigations, various researchers have highlighted the presence of pseudogenes and the phenomenon of mitochondrial introgression (Kemppainen *et al.*, 2009).

Exploring reproductive isolation in marine ecosystems presents significant challenges, as it constitutes a fundamental aspect of the biological species concept. To establish a relationship between the concept of biological species and DNA barcoding, a research study focusing on cosmopolitan marine bryozoans illustrated that distinct barcode clusters corresponded to populations reproductively isolated (Gomez *et al.*, 2007).

The process of finding novel species and recognizing those already known will be enhanced by incorporating morphological, environmental, and functional information in conjunction with DNA barcode data (Padial *et al.*, 2010). It has been identified that numerous inadequacies exist in the integration of DNA barcode data (DeSalle *et al.*, 2005).

6.11 ADVANCEMENTS IN DNA BARCODING TECHNIQUES

Given that merely 48% of marine species have been identified thus far, diverse arrays of DNA barcode research initiatives are underway and are anticipated to expand shortly. We must prioritize the advancement of innovative methodologies for the collection of samples from the deep sea, alongside the formulation of effective strategies for their preservation. Integrating laboratories into the process presents a significant advantage, as it facilitates early sampling and mitigates the risk of DNA leakage, thereby ensuring precise DNA amplification outcomes. The advancement of DNA microarrays, commonly called chips, for an extensive array of marine species is undoubtedly feasible (Kochzius *et al.*, 2008)

Merging genomic research with DNA barcoding offers a prevailing key to tackling the challenges that arise after applying DNA barcoding techniques. The screening of cDNA significantly possesses the capacity to mitigate errors that arise from inadequate preparation methods. A substantial volume of data will be essential for accurate identifications and analyses in the foreseeable future. It occurs notwithstanding the presence of numerous advantages, as well as limitations, that are inherently linked to this technology (Trivedi *et al.*, 2014)

The DNA barcoding initiative has largely driven recent advancements in ecosystem and species biology. Identifying organisms in stomach extracts significantly enhances our comprehension of the dietary habits of wild animals, especially in scenarios in which behavioral studies may not be feasible (Passmore *et al.*, 2006). DNA barcoding can effectively clarify the intricate relationships between hosts and their parasites, and those involving symbiotic interactions (Besansky *et al.*, 2003).

The methodology elucidates the transmission pathways of symbionts and parasites across successive generations of hosts. It is exemplified through the intricate interactions observed between beetles belonging to the Lecythidaceae family and their associated endosymbiotic yeasts, including various clades of *Candida* spp., as well as other unidentified yeast species (Berkov *et al.*, 2007). The molecular chronology of relations like symbiosis is construed through barcoding techniques (Anker *et al.*, 2007)

Innovative sequencing methodologies, such as pyrosequencing (including 454, Solexa, and SOLID), provide fast and comprehensive analysis of diverse samples, including, but not limited to, stomach contents, water columns, blood, and food. This advancement, primarily employed in the rapidly evolving domain of metagenomics, presents significant potential for forthcoming DNA barcoding initiatives (Hudson, 2008). DNA barcoding is a massive technological advancement. One essential criterion for submitting data to BOLD is the maintenance of comprehensive structural references for species, commonly referred to as vouchers. Innovative methodologies for non-destructive DNA extraction from freshly acquired specimens have been formulated (Rowley *et al.*, 2007).

A major constraint of molecular-based investigations, including DNA barcoding, is the challenge of extracting DNA from specimens preserved in formalin. The collections of animals housed within museums serve as the principal repository of voucher specimens utilized for species descriptions, with a significant proportion of these specimens being preserved in formalin. The central challenge lies in determining appropriate methodologies for extracting DNA from formalin-preserved specimens and acquiring DNA barcodes from these materials (Frézal and Leblois, 2008).

6.11.1 Next-Generation Sequencing and Metabarcoding

A groundbreaking advancement has emerged in high-throughput DNA sequencing technology, which has tremendous potential for deepening our comprehension of marine nematodes' diversity and roles in ecological processes across diverse geographical contexts. Including Roche 454, Illumina (HiSeq and Mseq), and ABI SOLiD, biological research has experienced a revolution. It has been made possible by the ability of these platforms to handle sequences of millions in a single run, which has helped improve the development of applications in marine ecosystems (Bik *et al.*, 2012).

Most of the techniques that next-generation sequencing (NGS) has entered have mainly taken a pilot using 18S rRNA as the molecular marker. The regions chosen for this study are the same as in the previous studies conducted on marine nematodes using DNA barcoding techniques. A systematic review approach using NGS has been utilized to determine the variety of marine nematodes living along the

UK coast (Trivedi *et al.*, 2014). Integrating genomic research with DNA barcoding is considered one of the most effective methods for assessing global biodiversity. Established in 2004, the Centre for Biodiversity Genomics (CBG) at the University of Guelph, Canada, has been a leading hub for DNA barcoding on an international scale. Identifying high-throughput sequencing technology opens great opportunities for its application in the DNA barcoding of massive amounts of environmental sources. The assessment of the samples from the quest for zooplankton in the Equatorial Pacific Ocean confirmed the high efficiency of the method of environmental barcoding as a conceptual direction (Machida *et al.*, 2009).

This technique, known as cDNA library screening, offers great potential in circumventing the artifacts resulting from problems such as mispriming or pseudogenes. Real-time qPCR appears valuable in providing a good prognosis for estimating species diversity and relative abundance, particularly of pooled and heterogeneous samples. It is hereby proposed that a taxonomic method that involves both morphological characteristics and DNA sequences be used to identify species to help optimize the process. The possibilities of DNA barcoding application for the estimation, investigation, and conservation of marine biodiversity are impressive, especially due to the availability of more effective and modern methodologies (Trivedi *et al.*, 2016a).

6.12 FUTURE PROSPECTS OF DNA BARCODING IN MARINE RESEARCH

Species detection and documentation of biodiversity consist of several vital stages, one of which is species determination. The identification process has always involved morphological analysis, which is owed to taxonomic research. There is growing concern about biodiversity among the several fields of study, including ecology, evolution, agriculture, and the economy. As a result, species identification, or determination of the species present, has become crucial. However, it has been noted that the number of specialists, such as taxonomists and others, participating in the identification processes has significantly decreased. Consequently, non-professionals need to possess alternative forms of accurate identification. Utilizing molecular data instead of morphological data to identify taxa represents one of the most promising methodologies in contemporary research. This topic has occupied the thoughts of numerous biologists for an extended period and is regarded as a foundational concept within the field (Pasha *et al.*, 2016).

Two main advancements in DNA barcoding include the progress of methods for sequencing mixed samples and the implementation of high-throughput sequencing techniques for PCR products using universal primers, commonly called DNA metabarcoding. Both of these developments possess the capacity to enhance the discipline markedly. The metabarcoding approach facilitates the thorough analysis of whole communities by examining diverse and intricate samples (Beng and Corlett, 2020).

Species identification is revolutionized with DNA barcoding by offering a rapid, standardized, molecular, and reliable method. DNA barcoding has revolutionized species identification by providing a fast, reliable, standardized molecular method.

Its application in marine biodiversity studies has yielded critical insights into species diversity, distribution, and evolutionary relationships, particularly in cryptic and morphologically ambiguous taxa. Despite challenges, such as gaps in barcode libraries and technical limitations, ongoing advancements in sequencing technologies and bioinformatics tools enhance its precision and accessibility. The integration of DNA barcoding with ecological and conservation initiatives underscores its significance in addressing global biodiversity challenges, such as safeguarding endangered species and measuring the impacts of environmental changes on marine ecosystems through DNA analysis. Barcoding thus stands as a cornerstone of modern biodiversity science, with promising prospects for future research and applications.

REFERENCES

Alfonsi, E., E. Méheust, S. Fuchs, F.-G. Carpentier, Y. Quillivic, A. Viricel, S. Hassani and J.-L. Jung. 2013. The use of DNA barcoding to monitor the marine mammal biodiversity along the French Atlantic coast. *ZooKeys*. 5.

Aliabadian, M., K.K. Beentjes, C.K. Roselaar, H. Van Brandwijk, V. Nijman and R. Vonk. 2013. DNA barcoding of Dutch birds. *Zookeys*. 25.

Andriollo, T. and M. Ruedi. 2020. Novel molecular tools to identify Plecotus bats in sympatry and a review of their distribution in Switzerland. *Revue suisse de Zoologie*. 125:61–72.

Anker, A., C. Hurt and N. Knowlton. 2007. Revision of the Alpheus nuttingi (Schmitt) species complex (Crustacea: Decapoda: Alpheidae), with description of a new species from the tropical eastern Pacific. *Zootaxa*. 1577:41–60.

Antil, S., J.S. Abraham, S. Sripoorna, S. Maurya, J. Dagar, S. Makhija, P. Bhagat, R. Gupta, U. Sood and R. Lal. 2023. DNA barcoding, an effective tool for species identification: a review. *Molecular Biology Reports*. 50:761–775.

Beng, K.C. and R.T. Corlett. 2020. Applications of environmental DNA (eDNA) in ecology and conservation: opportunities, challenges and prospects. *Biodiversity and Conservation*. 29:2089–2121.

Berkov, A., J. Feinstein, J. Small, M. Nkamany and P. Centeno. 2007. Yeasts isolated from neotropical wood-boring beetles in SE Peru. *Biotropica*. 39:530–538.

Besansky, N.J., D.W. Severson and M.T. Ferdig. 2003. DNA barcoding of parasites and invertebrate disease vectors: what you don't know can hurt you. *Trends in Parasitology*. 19:545–546.

Bhadury, P., M.C. Austen, D.T. Bilton, P.J.D. Lambshead, A.D. Rogers and G.R. Smerdon. 2006. Development and evaluation of a DNA-barcoding approach for the rapid identification of nematodes. *Marine Ecology Progress Series*. 320:1–9.

Bhattacharjee, D., B. Samanta, A.A. Danda and P. Bhadury. 2013. Temporal succession of phytoplankton assemblages in a tidal creek system of the Sundarbans mangroves: an integrated approach. *International Journal of Biodiversity*. 2013:824543.

Bik, H.M., D.L. Porazinska, S. Creer, J.G. Caporaso, R. Knight and W.K. Thomas. 2012. Sequencing our way towards understanding global eukaryotic biodiversity. *Trends in Ecology & Evolution*. 27:233–243.

Blanco-Bercial, L., F. Álvarez-Marqués and A. Bucklin. 2011. Comparative phylogeography and connectivity of sibling species of the marine copepod Clausocalanus (Calanoida). *Journal of Experimental Marine Biology and Ecology*. 404:108–115.

Boeuf, G. 2011. Marine biodiversity characteristics. *Comptes Rendus Biologies*. 334:435–440.

Bouchet, P. 2006. The magnitude of marine biodiversity. *The exploration of marine biodiversity: scientific and technological challenges*. 31–62. Fundación BBVA, Bilbao, Spain.

Briggs, J.C. 1994. Species diversity: land and sea compared. *Systematic Biology*. 43:130–135.

Bucklin, A., R.R. Hopcroft, K.N. Kosobokova, L.M. Nigro, B.D. Ortman, R.M. Jennings and C.J. Sweetman. 2010. DNA barcoding of Arctic Ocean holozooplankton for species identification and recognition. *Deep Sea Research Part II: Topical Studies in Oceanography*. 57:40–48.

Chen, S., H. Yao, J. Han, C. Liu, J. Song, L. Shi, Y. Zhu, X. Ma, T. Gao and X. Pang. 2010. Validation of the ITS2 region as a novel DNA barcode for identifying medicinal plant species. *PLoS One*. 5:e8613.

Daru, B.H., K. Yessoufou, L.T. Mankga and T.J. Davies. 2013. A global trend towards the loss of evolutionarily unique species in mangrove ecosystems. *PLoS One*. 8:e66686.

DeLong, D.W. 2004. *Lost knowledge: Confronting the threat of an aging workforce*. Oxford University Press.

DeSalle, R., M.G. Egan and M. Siddall. 2005. The unholy trinity: taxonomy, species delimitation and DNA barcoding. *Philosophical Transactions of the Royal Society B: Biological Sciences*. 360:1905–1916.

DeSalle, R. and P. Goldstein. 2019. Review and interpretation of trends in DNA barcoding. *Frontiers in Ecology and Evolution*. 7:302.

De Wit, P., E. Rota and C. Erseus. 2009. Grania (Annelida: Clitellata: Enchytraeidae) of the Great Barrier Reef, Australia, including four new species and a re-description of Grania trichaeta Jamieson, 1977. *Zootaxa*. 2165:16–38.

Dhaneesh, K.V., T.T.A. Kumar and A.B. Kumar. 2015. Barcoding, phylogeography and species boundaries in clownfishes of the Indian Ocean. *DNA Barcodes*. 3:5–16.

Efe, M.A., E.S. Tavares, A.J. Baker and S.L. Bonatto. 2009. Multigene phylogeny and DNA barcoding indicate that the sandwich tern complex (Thalasseus sandvicensis, Laridae, Sternini) comprises two species. *Molecular Phylogenetics and Evolution*. 52:263–267.

Erpenbeck, D., J.N. Hooper and G. Wörheide. 2006. CO1 phylogenies in diploblasts and the 'barcoding of life'—are we sequencing a suboptimal partition? *Molecular Ecology Notes*. 6:550–553.

Evans-Illidge, E., N. Andreakis, L. Hoj, P. Kearns, M.R. Hall, G. Ericson, R.E. Cobb and B.R. Gordon. 2015. Diversity of marine-derived fungal cultures exposed by DNA barcodes: the algorithm matters.

Fišer Pečnikar, Ž. and E.V. Buzan. 2014. 20 years since the introduction of DNA barcoding: from theory to application. *Journal of Applied Genetics*. 55:43–52.

Frézal, L. and R. Leblois. 2008. Four years of DNA barcoding: current advances and prospects. *Infection, Genetics and Evolution*. 8:727–736.

Gomez, A., P.J. Wright, D.H. Lunt, J.M. Cancino, G.R. Carvalho and R.N. Hughes. 2007. Mating trials validate the use of DNA barcoding to reveal cryptic speciation of a marine bryozoan taxon. *Proceedings of the Royal Society B: Biological Sciences*. 274:199–207.

Gong, S., Y. Ding, Y. Wang, G. Jiang and C. Zhu. 2018. Advances in DNA barcoding of toxic marine organisms. *International Journal of Molecular Sciences*. 19:2931.

Hajibabaei, M., M.A. Smith, D.H. Janzen, J.J. Rodriguez, J.B. Whitfield and P.D. Hebert. 2006. A minimalist barcode can identify a specimen whose DNA is degraded. *Molecular Ecology Notes*. 6:959–964.

Hebert, P.D., A. Cywinska, S.L. Ball and J.R. Dewaard. 2003. Biological identifications through DNA barcodes. *Proceedings of the Royal Society of London. Series B: Biological Sciences*. 270:313–321.

Heinrichs, J., H.P. Kreier, K. Feldberg, A.R. Schmidt, R.L. Zhu, B. Shaw, A.J. Shaw and V. Wissemann. 2011. Formalizing morphologically cryptic biological entities: new insights from DNA taxonomy, hybridization, and biogeography in the leafy liverwort Porella platyphylla (Jungermanniopsida, Porellales). *American Journal of Botany*. 98:1252–1262.

Hirose, M. and E. Hirose. 2009. DNA barcoding in photosymbiotic species of diplosoma (Ascidiacea: Didemnidae), with the description of a new species from the Southern Ryukyus, Japan. *Zoological Science*. 26:564–568.

Holmes, B.H., D. Steinke and R.D. Ward. 2009. Identification of shark and ray fins using DNA barcoding. *Fisheries Research*. 95:280–288.

Hudson, M.E. 2008. Sequencing breakthroughs for genomic ecology and evolutionary biology. *Molecular Ecology Resources*. 8:3–17.

Jena, B., A.K. Giri, B. Biswal, L. Acharya and M.K. Panda. 2019. Cultivar identification in Tabernaemontana divaricata (L.) R. Br. ex Roem. & Schult. using universal barcode markers. *Gene Reports*. 17:100467.

Jennings, R.M., A. Bucklin and A. Pierrot-Bults. 2010. Barcoding of arrow worms (Phylum Chaetognatha) from three oceans: genetic diversity and evolution within an enigmatic phylum. *PLoS One*. 5:e9949.

Ji, Y., T. Huotari, T. Roslin, N.M. Schmidt, J. Wang, D.W. Yu and O. Ovaskainen. 2020. SPIKEPIPE: a metagenomic pipeline for the accurate quantification of eukaryotic species occurrences and intraspecific abundance change using DNA barcodes or mitogenomes. *Molecular Ecology Resources*. 20:256–267.

Kamaruzzaman, B., B.A. John, K. Zaleha and K. Jalal. 2011. Molecular phylogeny of horseshoe crab. *Asian Journal of Biotechnology*. 3:302–309.

Kaur, S. 2015. DNA barcoding and its applications. *International Journal of Engineering Research and General Science*. 3:602–604.

Kemppainen, P., M. Panova, J. Hollander and K. Johannesson. 2009. Complete lack of mitochondrial divergence between two species of NE Atlantic marine intertidal gastropods. *Journal of Evolutionary Biology*. 22:2000–2011.

Keshavmurthy, S., S.-Y. Yang, A. Alamaru, Y.-Y. Chuang, M. Pichon, D. Obura, S. Fontana, S. De Palmas, F. Stefani and F. Benzoni. 2013. DNA barcoding reveals the coral "laboratory-rat", Stylophora pistillata encompasses multiple identities. *Scientific Reports*. 3:1520.

Kim, M.S., M.Y. Yang and G.Y. Cho. 2010. Applying DNA barcoding to Korean Gracilariaceae (Rhodophyta). *Cryptogamie Algologie*. 31:387.

Kochzius, M., M. Nölte, H. Weber, N. Silkenbeumer, S. Hjörleifsdottir, G. Hreggvidsson, V. Marteinsson, K. Kappel, S. Planes and F. Tinti. 2008. DNA microarrays for identifying fishes. *Marine Biotechnology*. 10:207–217.

Kress, W.J. and D.L. Erickson. 2008. DNA barcodes: genes, genomics, and bioinformatics. *Proceedings of the National Academy of Sciences*. 105:2761–2762.

Kress, W.J., K.J. Wurdack, E.A. Zimmer, L.A. Weigt and D.H. Janzen. 2005. Use of DNA barcodes to identify flowering plants. *Proceedings of the National Academy of Sciences*. 102:8369–8374.

Lahaye, R., M. Van Der Bank, D. Bogarin, J. Warner, F. Pupulin, G. Gigot, O. Maurin, S. Duthoit, T.G. Barraclough and V. Savolainen. 2008. DNA barcoding the floras of biodiversity hotspots. *Proceedings of the National Academy of Sciences*. 105:2923–2928.

Lambert, D., A. Baker, L. Huynen, O. Haddrath, P. Hebert and C. Millar. 2005. Is a large-scale DNA-based inventory of ancient life possible? *Journal of Heredity*. 96:279–284.

Layton, K.K., A.L. Martel and P.D. Hebert. 2014. Patterns of DNA barcode variation in Canadian marine molluscs. *PLoS One*. 9:e95003.

Lee, G., S.Y. Park, J. Hwang, Y.-H. Lee, S.Y. Hwang, S. Lee and T.-K. Lee. 2011. Development of DNA chip for jellyfish verification from South Korea. *Biochip Journal*. 5:375–382.

Li, X., Y. Yang, R.J. Henry, M. Rossetto, Y. Wang and S. Chen. 2015. Plant DNA barcoding: from gene to genome. *Biological Reviews*. 90:157–166.

Lobo, J., M.A. Teixeira, L.M. Borges, M.S. Ferreira, C. Hollatz, P.T. Gomes, R. Sousa, A. Ravara, M.H. Costa and F.O. Costa. 2016. Starting a DNA barcode reference library for shallow water polychaetes from the Southern European Atlantic coast. *Molecular Ecology Resources*. 16:298–313.

Lucas, C., T. Thangaradjou and J. Papenbrock. 2012. Development of a DNA barcoding system for seagrasses: successful but not simple. *PLoS One*. 7:e29987.

Maas, A.E., L. Blanco-Bercial and G.L. Lawson. 2013. Reexamination of the species assignment of Diacavolinia pteropods using DNA barcoding. *PLoS One*. 8:e53889.

Machida, R.J., Y. Hashiguchi, M. Nishida and S. Nishida. 2009. Zooplankton diversity analysis through single-gene sequencing of a community sample. *BMC Genomics*. 10:1–7.

Machida, R.J. and A. Tsuda. 2010. Dissimilarity of species and forms of planktonic Neocalanus copepods using mitochondrial COI, 12S, nuclear ITS, and 28S gene sequences. *PLoS One*. 5:e10278.

Meier, R., K. Shiyang, G. Vaidya and P.K. Ng. 2006. DNA barcoding and taxonomy in Diptera: a tale of high intraspecific variability and low identification success. *Systematic Biology*. 55:715–728.

Mitra, A., S. Trivedi, S. Zaman, P. Pramanick, S. Chakraborty, N. Pal, P. Fazli and K. Banerjee. 2015. Decadal variation of nutrient level in two major estuaries in Indian Sundarbans. *Jordan Journal of Biological Sciences*. 147:1–6.

Molnar, J.L., R.L. Gamboa, C. Revenga and M.D. Spalding. 2008. Assessing the global threat of invasive species to marine biodiversity. *Frontiers in Ecology and the Environment*. 6:485–492.

Nagy, Z.T., G. Sonet, F. Glaw and M. Vences. 2012. First large-scale DNA barcoding assessment of reptiles in the biodiversity hotspot of Madagascar, based on newly designed COI primers. *PLoS One*. 7:e34506.

Naro-Maciel, E., B. Reid, N.N. Fitzsimmons, M. Le, R. Desalle and G. Amato. 2010. DNA barcodes for globally threatened marine turtles: a registry approach to documenting biodiversity. *Molecular Ecology Resources*. 10:252–263.

O'Brien, H.E., J.L. Parrent, J.A. Jackson, J.-M. Moncalvo and R. Vilgalys. 2005. Fungal community analysis by large-scale sequencing of environmental samples. *Applied and Environmental Microbiology*. 71:5544–5550.

Padial, J.M., A. Miralles, I. De La Riva and M. Vences. 2010. The integrative future of taxonomy. *Frontiers in Zoology*. 7:1–14.

Park, M.-H., J.-H. Jung, E. Jo, K.-M. Park, Y.-S. Baek, S.-J. Kim and G.-S. Min. 2019. Utility of mitochondrial CO1 sequences for species discrimination of Spirotrichea ciliates (Protozoa, Ciliophora). *Mitochondrial DNA Part A*. 30:148–155.

Pasha, F., S. Saggu and M.F. Albalawi. 2016. Future perspectives of DNA barcoding in marine zooplanktons and invertebrates. In *DNA Barcoding in Marine Perspectives: Assessment and Conservation of Biodiversity*, pp. 169–179. Cham: Springer International Publishing.

Passmore, A., S. Jarman, K. Swadling, S. Kawaguchi, A. McMinn and S. Nicol. 2006. DNA as a dietary biomarker in Antarctic krill, Euphausia superba. *Marine Biotechnology*. 8:686–696.

Pauls, S.U., R.J. Blahnik, X. Zhou, C.T. Wardwell and R.W. Holzenthal. 2010. DNA barcode data confirm new species and reveal cryptic diversity in Chilean Smicridea (Smicridea) (Trichoptera: Hydropsychidae). *Journal of the North American Benthological Society*. 29:1058–1074.

Planes, S., M.J. Feio, A.F. Filipe, A. Garcia-Raventós, A. Ardura, A.R. Calapez, A.M. Pujante, A. Mortágua, C. Múrria and D. Diaz-De-Quijano. 2020. *Advances in the use of molecular tools in ecological and biodiversity assessment of aquatic ecosystems*. Congress of the Iberian-Association-of-Limnology (AIL). 19.

Punina, E., E. Machs, E. Krapivskaya and A. Rodionov. 2017. Polymorphic sites in transcribed spacers of 35S rRNA genes as an indicator of origin of the Paeonia cultivars. *Russian Journal of Genetics*. 53:202–212.

Purty, R. and S. Chatterjee. 2016. DNA barcoding: an effective technique in molecular taxonomy. *Austin Journal of Biotechnology & Bioengineering*. 3:1059.

Razy-Krajka, F., K. Lam, W. Wang, A. Stolfi, M. Joly, R. Bonneau and L. Christiaen. 2014. Collier/OLF/EBF-dependent transcriptional dynamics control pharyngeal muscle specification from primed cardiopharyngeal progenitors. *Developmental Cell*. 29:263–276.

Rico-Sánchez, A.E., A. Sundermann, E. López-López, M.J. Torres-Olvera, S.A. Mueller and P.J. Haubrock. 2020. Biological diversity in protected areas: not yet known but already threatened. *Global Ecology and Conservation*. 22:e01006.

Rowley, D.L., J.A. Coddington, M.W. Gates, A.L. Norrbom, R.A. Ochoa, N.J. Vandenberg and M.H. Greenstone. 2007. Vouchering DNA-barcoded specimens: test of a nondestructive extraction protocol for terrestrial arthropods. *Molecular Ecology Notes*. 7:915–924.

Rubinoff, D., S. Cameron and K. Will. 2006. A genomic perspective on the shortcomings of mitochondrial DNA for "barcoding" identification. *Journal of Heredity*. 97: 581–594.

Schindel, D.E. and S.E. Miller. 2005. DNA barcoding a useful tool for taxonomists. *Nature*. 435(7038):17–17.

Schindel, D.E., M.Y. Stoeckle, C. Milensky, M. Trizna, B. Schmidt, C. Gebhard and G. Graves. 2011. Project description: DNA barcodes of bird species in the national museum of natural history, Smithsonian Institution, USA. *ZooKeys*. 87.

Schuster, A., D. Erpenbeck, A. Pisera, J. Hooper, M. Bryce, J. Fromont and G. Wörheide. 2015. Deceptive desmas: molecular phylogenetics suggests a new classification and uncovers convergent evolution of lithistid demosponges. *PLoS One*. 10:e116038.

Shadrin, D. 2021. DNA barcoding: applications. *Russian Journal of Genetics*. 57:489–497.

Sire, L., D. Gey, R. Debruyne, T. Noblecourt, F. Soldati, T. Barnouin, G. Parmain, C. Bouget, C. Lopez-Vaamonde and R. Rougerie. 2019. The challenge of DNA barcoding saproxylic beetles in natural history collections—exploring the potential of parallel multiplex sequencing with Illumina MiSeq. *Frontiers in Ecology and Evolution*. 7:495.

Smith, K.F., J. Thia, C.E. Gemmill, S.C. Cary and A.E. Fidler. 2012. Barcoding of the cytochrome oxidase I (COI) indicates a recent introduction of Ciona savignyi into New Zealand and provides a rapid method for Ciona species discrimination. *Aquatic Invasions*. 7:305–313.

Sperling, E.A., R.D. Rosengarten, M.A. Moreno and S.L. Dellaporta. 2012. The complete mitochondrial genome of the verongid sponge Aplysina cauliformis: implications for DNA barcoding in demosponges. In *Ancient Animals, New Challenges: Developments in Sponge Research,* pp. 61–69. Dordrecht: Springer Netherlands.

Tack, J.F., E. Vanden Berghe and P. Polk. 1992. *Ecomorphology of Crassostrea cucullata (Born, 1778)(Ostreidae) in a mangrove creek (Gazi, Kenya)*. The Ecology of Mangrove and Related Ecosystems: Proceedings of the International Symposium held at Mombasa, Kenya, 24–30 September 1990, Springer. 109–117.

Trivedi, S., A.A. Aloufi, A.A. Ansari and S.K. Ghosh. 2015. Molecular phylogeny of oysters belonging to the genus Crassostrea through DNA barcoding. *Journal of Entomology and Zoology Studies* 3:21–26.

Trivedi, S., A.A. Aloufi, A.A. Ansari and S.K. Ghosh. 2016a. Role of DNA barcoding in marine biodiversity assessment and conservation: an update. *Saudi Journal of Biological Sciences*. 23:161–171.

Trivedi, S., A.A. Aloufi, H. Rehman, S. Saggu and S.K. Ghosh. 2016b. DNA barcoding: tool for assessing species identification in Reptilia. *Journal of Entomology and Zoology Studies*. 4:332–337.

Trivedi, S., A.A. Ansari, S.K. Ghosh and H. Rehman. 2014. *DNA barcoding in marine perspectives*. Aqaba International Conference on Marine and Coastal Environment, Status and Challenges in Arab World, Aqaba, Jordan, Springer.

Trivedi, S., A.A. Ansari, H. Rehman, S. Saggu, Z. Abbas and S. Ghosh. 2016c. DNA barcoding as a molecular tool for the assessment of plant biodiversity. *Plant biodiversity: monitoring assessment and conservation*. CABI, Wallingford. 419–428.

Uthicke, S., M. Byrne and C. Conand. 2010. Genetic barcoding of commercial Bêche-de-mer species (Echinodermata: Holothuroidea). *Molecular Ecology Resources*. 10:634–646.

Vanhove, M.P., B. Tessens, C. Schoelinck, U. Jondelius, D.T.J. Littlewood, T. Artois and T. Huyse. 2013. Problematic barcoding in flatworms: a case-study on monogeneans and rhabdocoels (Platyhelminthes). *ZooKeys*. 355.

Vargas, S., M. Kelly, K. Schnabel, S. Mills, D. Bowden and G. Woerheide. 2015. Diversity in a cold hot-spot: DNA-barcoding reveals patterns of evolution among Antarctic Demosponges (class Demospongiae, phylum Porifera). *PLoS One*. 10:e0127573.

Vargas, S.M., F.C. Araújo and F.R. Santos. 2009. DNA barcoding of Brazilian sea turtles (Testudines). *Genetics and Molecular Biology*. 32:608–612.

Velmurugan, S., C. Prasannakumar, S. Manokaran, T.A. Kumar, A. Samkamaleson and A. Palavesam. 2013. DNA barcodes for marine fungal identification and discovery. *Fungal Ecology*. 6:408–418.

Vences, M., M. Thomas, A. Van Der Meijden, Y. Chiari and D.R. Vieites. 2005. Comparative performance of the 16S rRNA gene in DNA barcoding of amphibians. *Frontiers in Zoology*. 2:1–12.

Ward, R.D., R. Hanner and P.D. Hebert. 2009. The campaign to DNA barcode all fishes, FISH-BOL. *Journal of Fish Biology*. 74:329–356.

Ward, R.D., B.H. Holmes and T.D. O'Hara. 2008. DNA barcoding discriminates echinoderm species. *Molecular Ecology Resources*. 8:1202–1211.

Waycott, M., C.M. Duarte, T.J. Carruthers, R.J. Orth, W.C. Dennison, S. Olyarnik, A. Calladine, J.W. Fourqurean, K.L. Heck Jr and A.R. Hughes. 2009. Accelerating loss of seagrasses across the globe threatens coastal ecosystems. *Proceedings of the National Academy of Sciences*. 106:12377–12381.

Waycott, M., D.W. Freshwater, R.A. York, A. Calladine and W.J. Kenworthy. 2002. Evolutionary trends in the seagrass genus Halophila (Thouars): insights from molecular phylogeny. *Bulletin of Marine Science*. 71:1299–1308.

Weigt, L.A., C.C. Baldwin, A. Driskell, D.G. Smith, A. Ormos and E.A. Reyier. 2012. Using DNA barcoding to assess Caribbean reef fish biodiversity: expanding taxonomic and geographic coverage. *PLoS One*. 7:e41059.

Wirth, T., D. Falush, R. Lan, F. Colles, P. Mensa, L.H. Wieler, H. Karch, P.R. Reeves, M.C. Maiden and H. Ochman. 2006. Sex and virulence in Escherichia coli: an evolutionary perspective. *Molecular Microbiology*. 60:1136–1151.

Zemlak, T.S., R.D. Ward, A.D. Connell, B.H. Holmes and P.D. Hebert. 2009. DNA barcoding reveals overlooked marine fishes. *Molecular Ecology Resources*. 9:237–242.

Zhang, H., Y. Zhang, Z. Zhang and T. Gao. 2013. DNA barcodes of eight species in genus Sebastes. *Biochemical Systematics and Ecology*. 48:45–50.

Zhao, X., S. Pang, T. Shan and F. Liu. 2013. Applications of three DNA barcodes in assorting intertidal red macroalgal flora in Qingdao, China. *Journal of Ocean University of China*. 12:139–145.

Zhou, J., M.E. Davey, J.B. Figueras, E. Rivkina, D. Gilichinsky and J.M. Tiedje. 1997. Phylogenetic diversity of a bacterial community determined from Siberian tundra soil DNA. *Microbiology*. 143:3913–3919.

7 DNA Barcoding in Animals for Integrative Taxonomy and Biodiversity

Humaira Muzaffar, Ghulam Hussain, Arslan Iftikhar, Misbah Ijaz, Muhammad Tahir Mohyud-din, Faiza Hassan, Misbah Batool, Shawana Iqbal, Ansha Ghaffar, Muhammad Mushahid, and Mahrukh Babar

7.1 INTRODUCTION

Various reliable molecular techniques are being developed rapidly to recognize animal and plant species. DNA barcoding is one such technique based on PCR genotyping and has been used extensively to recognize plant and animal species efficiently (Chen et al., 2021). DNA barcoding is recognized as an innovative technique for quick and precise recognition of species using DNA sequences obtained from minute tissue samples. This technique is currently utilized across a wide range of biological groups. In taxonomy, DNA barcoding refines species recognition by including every phase of an organism's life cycle. DNA barcoding as a tool for discovering biodiversity enables the detection of species that may be exceptional in science (Kress & Erickson, 2012).

The earth's biota may include between 10 and 100 million eukaryotic species. The recent advancement in DNA-sequencing-based technology, known as DNA barcoding, has made it possible to utilize tiny variants of DNA sequences as species identifiers. This technique is now being used extensively, especially by scientists studying organisms like protists, red algae (*Rhodophyta*), bacteria, and viruses that are difficult to identify on a morphological basis (Waugh, 2007). In recent years, DNA barcoding has emerged as an efficient and reliable technique capable of identifying already established species and discovering new ones by analyzing sequence variations in a standardized DNA segment (Yang et al., 2018). DNA barcoding identifies species based on variations in the conserved sequence of DNA. The term 'barcodes' denotes analyzing multiple samples by collectively sequencing identifier sequences. This method has been used to categorize various organisms, from the simplest to the most complex. In contrast to other marker analytical techniques, such as short tandem repeats (STR), which are specifically used to recognize individual organisms,

DOI: 10.1201/9781003538929-7

DNA barcoding focuses on species-level identification. The Barcode of Life Data Systems database, an online repository of reference DNA sequences for comparison with the DNA sequence of an unknown organism, is where barcode sequences can be accessed (Mir et al., 2021).

In 2003, researchers at the University of Guelph, Canada, introduced the groundbreaking concept of DNA barcoding. The gene encoding for mitochondrial cytochrome c oxidase subunit 1 (COI) was identified as a region specific to animals for use as a barcode (Hebert et al., 2003). The Universal Product Code's black stripes, which are used to identify commercial goods, are comparable to the DNA barcode. Genebanks already contain significant sequence data for the recognized barcoding loci. However, more data are needed in databases to utilize DNA barcodes properly. DNA barcoding is sought to be designated as a universal benchmark for the classification of biological organisms by the International Barcode of Life Consortium (iBOL), established in 2004. The primary goals involve enhancing taxonomy to advance science and society by recognizing known specimens and uncovering previously unknown entities (von Cräutlein et al., 2011).

An ideal DNA barcode should enable quick, reliable, automated, and cost-effective species identification, even for users with minimal taxonomic expertise (Luo et al., 2011). DNA barcoding allows recognizing, cataloging, and examining specimens to (i) comprehend species variety in an environment, and (ii) assess genetic heterogeneity within species. More precisely, DNA barcodes offer valuable information to various conservation stakeholders: researchers can swiftly identify species, taxonomists can identify species groups that need more in-depth research, and policymakers can utilize barcode data to determine the proper conservation dimensions (Krishna Krishnamurthy & Francis, 2012).

For a gene region to function effectively as a DNA barcode, it must fulfill three primary criteria: it should display significant genetic variability at the species level, contain conserved flanking sites suitable for universal PCR primers, and maintain an appropriate length for the current capabilities of DNA extraction and sequencing. Furthermore, some researchers also regard the sequence quality as important (Kress & Erickson, 2012).

The mitochondrial gene encoding for the enzyme cytochrome c oxidase subunit 1 (CO1) satisfies these requirements and is extensively utilized as a standardized species-level barcode for numerous animals. Nevertheless, CO1 is ineffective for plants and fungi (Chase et al., 2005), motivating botanists to pursue more appropriate markers. In 2008, the Consortium for the Barcode of Life working for plants evaluated multiple candidate gene regions and identified three key markers for plants: *trnH-psbA*, *matK*, and *rbcL*. *MatK* and *rbcL* were designated core barcodes, while *trnH-psbA* was recommended as an important supplementary marker (Pang et al., 2012). Moreover, some researchers continue to advocate additional markers, such as ITS, for specialized applications or specific taxa.

7.2 CHOICE OF BARCODE LOCI

When DNA barcoding was proposed, a section of the mitochondrial CO1 gene was chosen for animal identification. Later pilot studies have shown its extensive applicability in animal systems. Although the exact taxonomic bounds of CO1 barcoding in animals are unknown, most testing groups have found it effective for species discrimination. Among other eukaryotes, the selection of a DNA region suitable for barcoding has not received much attention, but among rRNA genes, prokaryotes are used for identification (Kress et al., 2005).

Over the past two decades, several major single-locus candidate barcodes have been proposed, including *psbA-trnH*, *rbcL*, *matK*, ITS2, ITS, and *trnL-trnF*. These barcodes have demonstrated high reliability in differentiating plant species (Zhu et al., 2022). In fungi, the internal transcribed spacer (ITS) gene exhibits a superior ability to distinguish closely related species compared to the CO1 gene (Imtiaz et al., 2017). The 16S rRNA gene is generally recognized as a key marker for identifying bacteria through DNA barcoding (Janda & Abbott, 2007). A two-step barcoding method is recommended for the identification of protists: the first step is the pre-barcoding step, which involves examining the V4 region of rDNA, while in the second step, additional barcodes designed for specific protest clades are used (Will et al., 2005). Generally, two main steps are involved in DNA barcoding: the first step involves the establishment of a barcode library of identified species, and the second step involves the identification of an unidentified sample by aligning its barcode sequence to the existing barcode library (Kaur, 2015).

7.3 LABORATORY TECHNIQUES FOR DNA BARCODING

The first step in the methodology of DNA barcoding is sample collection. The collection of samples depends on the requirements of the experiment. Samples may contain both fresh and processed materials. After collection, samples are preserved in a suitable preservative to maintain their structural integrity. The samples may be preserved in 4% formalin. However, preserving samples in formaldehyde forms formaldehyde–DNA complexes, facilitating the cross-links between protein and DNA, and possibly leading to DNA fragmentation. Thus, formaldehyde may impede the process of DNA extraction (Nakahama et al., 2019). Therefore, the samples are commonly stored in ethanol or using a cryopreservation technique to avoid these restrictions. After that, DNA is extracted from the samples. Various protocols exist for extracting DNA from microbes, animals, and plant tissue, including the phenol–chloroform method, cetyltrimethylammonium, and sodium dodecyl sulfate DNA extraction techniques (Ahmed, 2022). These protocols are performed by following a conventional approach to get an appropriate quantity of purified DNA that is free from metabolite contamination because the process of PCR amplification may be impeded by the contamination (Creer et al., 2010). After the samples' DNA data is sequenced, bioinformatics methods, relevant databases, and tools like BLAST and NCBI are used to begin the similarity-search process. To create a new barcode or validate a new record for an existing one, accurate data analysis is crucial (Yu et al., 2021).

A GenBank accession number is then requested for the sequence if the similarity search produces statistically significant results (Clark et al., 2016). Data from the NCBI is then incorporated within the Barcode of Life Database to determine species (Ratnasingham & Hebert, 2007). Nevertheless, research using DNA barcodes has grown rapidly, reaching nearly 1400 by 2020 (Elías-Gutiérrez & Valdez-Moreno, 2023). The approach has numerous applications because there is typically less variation among individuals of the same species than those of closely related species, though it is particularly valuable for species identification. Numerous species have been successfully identified by barcoding. There is usually a substantial degree of intraspecific K2P divergence in a sequence from a misdiagnosed species (Shen et al., 2013).

7.4 BIOINFORMATICS TOOLS AND DATABASES FOR DNA BARCODING

DNA barcoding involves several bioinformatics tools and databases to analyze and manage the genetic data. Here are some key resources:

7.4.1 Barcode of Life Database Systems (BOLD)

BOLD is recognized as a formidable repository for DNA barcoding (Ratnasingham & Hebert, 2007), where 8.5 million barcode records have been meticulously cataloged. Moreover, the vouchers of the respective specimens from which barcode sequences were derived are preserved carefully. The database includes comprehensive metadata, including museum vouchers, country of origin, and taxonomic details.

7.4.2 GenBank

The GenBank is a repository of DNA sequences curated by NCBI (Sayers et al., 2020). This repository contains a vast collection of barcode datasets. There is an outstanding surge in entries of DNA barcodes, particularly for COI genes (Porter & Hajibabaei, 2018).

7.4.3 Molecular Evolutionary Genetics Analysis (MEGA)

MEGA software performs a comparative analysis of protein and DNA sequences. This analytical software facilitates the reveal of the evolutionary trajectories at the molecular level across species, genes, and genomes over time. There are two different versions of MEGA software: the first version is known as the graphical user interface (MEGA-GUI), which offers visual tools for detailed data analysis and insightful interpretation of results (Tamura et al., 2011); the second version, known as the command-line edition (MEGA-CC), is designed and optimized precisely for integrated pipeline analysis (Kumar et al., 2012).

7.4.4 Multiple Sequence Comparison by Log-Expectation (MUSCLE)

Alignments of protein sequences are generated by the MUSCLE analytical tool (Williams et al., 2014). Alignments of multiple sequences have a broad range of applications, including the prediction of protein structure, essential residue identification, and constructing phylogenetic trees (Edgar, 2004).

7.4.5 Basic Local Alignment Search Tool (BLAST)

This search tool matches the sequence with GenBank. Its searches are very effective and widely used to confirm taxonomic classification (Erickson et al., 2008; Kõljalg et al., 2005).

7.5 APPLICATIONS OF DNA BARCODING

7.5.1 Species Identification

DNA barcoding is an analytical technique that allows the identification and classification of species by using a standardized segment of DNA known as barcodes (Ratnasingham & Hebert, 2013). The initial proposal of this technique was presented by Hebert et al. (2003), and it is considered a critical supplementary method to the conventional analytical procedures based on morphological analysis (Hollingsworth et al., 2016) for ecological forensics (Kesanakurti et al., 2011), compliance monitoring (Simberloff et al., 2013), and floristic and vegetation surveys (Parmentier et al., 2013).

7.5.2 Cryptic Species

Cryptic species are genetically distinct species classified previously as a single species or taxon because of their morphological similarities. However, DNA barcoding has made it possible to distinguish between these morphologically indistinguishable cryptic species using standardized genetic markers. So, this technique has been widely used to identify many novel species. Multiple research studies have confirmed the efficacy of DNA barcoding, with over 90% of analyzed species accurately differentiated at genetic and species levels (Mishra et al., 2016; Becker et al., 2011).

7.5.3 Molecular Phylogenetic

The main objective of DNA barcoding is the identification of species. Moreover, the evolutionary relationships have been deduced commonly at different levels by using mtDNA (DeSalle, 2007). This has prompted many researchers to inspect the phylogenetic utility of DNA barcode data. Some argue that DNA barcode data offers limited phylogenetic value due to its few informative characters. However, there is an ongoing debate about whether adding more species or genetic characters in phylogenetic studies is more advantageous. Many researchers have concluded that amalgamating more species can be equally, if not more, beneficial. For example, Hillis suggested that with limited time and resources, phylogenetic studies might benefit more from adding species rather than increasing the number of genetic characters analyzed.

7.5.4 Forensic Taxonomy

DNA barcoding is vital in forensic taxonomy, especially when dealing with crimes such as illegal trafficking and poaching. This technique helps in the enforcement of law and conservation of wildlife species by:

- Identification of species: DNA barcoding can help identify the species from which the processed or confiscated animal goods originated, thus helping in prosecutions.
- Tracking of source: DNA barcoding can also determine the geographic origin of the specimens. This will help identify areas with illegal activities (Odah, 2024).

7.5.5 Biodiversity Assessment and Conservation

The DNA barcoding technique has a significant role in conservation biology and ecosystem monitoring. It also helps conserve endangered species and identify invasive species, facilitating early intervention and control (Odah, 2024).

7.5.6 Quality Control in Biological Research

A massive volume of data from a growing number of labs has been deposited in GenBank. Aside from deliberate technological mistakes in sequencing, errors may also arise from sample mismatch, contamination, and inaccurate species identification. Inaccurate data can confuse taxonomies and phylogenies, which can significantly influence important forensic applications. These mistakes are frequently difficult to identify, and it happens much too often that neither the taxonomic accuracy nor the potential for contamination are confirmed. The dichotomy of tiny K2P interspecific differences and deep intraspecific differences implies that more accuracy verification is required. Of course, not all paradoxes result from contamination or incorrect species identification. Once exact genetic data are acquired for each species, especially from type localities, it becomes possible to precisely trace the origin of ambiguous sequences, whether they stem from sequencing inaccuracies or taxonomic misidentifications (Shen et al., 2013). The worldwide endeavor aims to catalog all amphibian and reptile species genetically (Murphy et al., 2013), making DNA barcoding invaluable for species authentication and detecting potential discrepancies in GenBank (Shen et al., 2013).

7.6 INTEGRATING DNA BARCODING WITH TRADITIONAL TAXONOMY

Traditionally, species differentiation relied on morphological features (Teodoro et al., 2016). However, identifying species based on morphology can be challenging, as distinctions among species are often subtle, particularly in comparisons of native versus non-native species (Pyšek et al., 2013). For the novice taxonomist, identifying even the smallest morphological variations requires a high skill level, which may eventually be necessary for identification consistency. Molecular technologies are being utilized increasingly to provide trustworthy techniques for identifying different plant and animal species (Chen et al., 2021). DNA barcoding, a PCR-based genotyping technique, has been widely used for this objective (Pawlowski et al., 2012). In traditional taxonomy, species classification was mainly based on morphological traits (Lim et al., 2019), making it very difficult and confusing to differentiate between cryptic species having morphological similarities (Galimberti et al., 2013). Conversely, DNA barcoding offers many advantages by providing rapid species identification (Antil et al., 2023), and objectivity in identifying species based on genetic variations (Gostel & Kress, 2022). The other benefits of DNA barcoding are listed below:

1. Cryptic Species Resolution: Many novel species have been revealed using the technique of DNA barcoding. These species were previously unknown because they could not be differentiated on a morphological basis only (Collins & Cruickshank, 2013).
2. Integration with Traditional Methods: Traditional taxonomy is designed to be augmented, not replaced, by DNA barcoding. Combining morphological traits with genetic data provides a more thorough method of separating species.
3. Applications Beyond Taxonomy: The adaptability of DNA barcoding is demonstrated by the fact that it can be used in forensics, ecological research, biodiversity assessment, and conservation (Gostel & Kress, 2022).

These advancements have improved our comprehension of the natural world, and they have established DNA barcoding as a vital instrument in modern taxonomy and biodiversity research (Odah, 2024).

7.7 CASE STUDIES IN ANIMAL TAXONOMY USING DNA BARCODING

The three main tenets of the DNA barcoding technique are scalability, minimalism, and standardization. In light of these guidelines, choosing appropriate barcoding entails identifying one or more standard loci for routine sequencing in a large and diverse group of samples, with the highest reliability to provide data sets that are easily comparable to differentiate across species. With 600–1000 base pair DNA sequences, the 5′-end of Cytochrome C Oxidase subunit one is generally thought to be a universal species-level barcode for animals, and it serves the function of accommodating interspecific heterogeneity (Kress & Erickson, 2012).

Variation in morphology and inadequate phylogenetic information owing to a lack of understanding of anatomical traits and the existence of cryptic species means that a morpho-molecular approach is required to improve identification accuracy and address current issues facing the animal kingdom (Wang et al., 2020). Uncovering phenotypic similarity and forecasting biodiversity responses to environmental changes require understanding and documentation of cryptic biodiversity (morphologically identical but genetically distinct species). A recent investigation in molecular systematics examining the amphibian *Limnonectes kuhlii* through mitochondrial DNA (mtDNA) analysis has uncovered 22 distinct evolutionary lineages. Among these, 16 are classified under *L. kuhlii*, a taxon once considered a single species. Frogs are widely acknowledged as a key group for uncovering cryptic species diversity. Because morphological changes do not necessarily indicate speciation, the underestimating of species diversity results from the underutilization of molecular methods (Mcleod et al., 2011). The emergence of cryptic and recently discovered species stresses the need for taxonomic revision of many animal taxa, which is highlighted by the DNA barcoding approach. DNA barcoding of 1,630 specimens has been used to examine the often-overlooked genus *Triplophysa* within the Qinghai-Tibet Plateau, a renowned hotspot for biodiversity. Twenty-four regional species, including two cryptic species, *T. robusta*, and *T. minxianensis*, were identified in this study (Wang et al., 2020). Moreover, there is an increasing inclination toward employing high-throughput technology for precise identification. When integrated with DNA metabarcoding, conventional trap-based monitoring strategies for controlling the spread of invasive insects become more efficient, as they facilitate the concurrent, multi-species detection of mixed populations (Piper et al., 2019).

Temperature sensitivity is predicted to increase in insects due to their short life cycles. For instance, significant poleward shifts have been recorded in non-migratory butterflies in Europe, where temperatures are a major influencing factor (Bale et al., 2002). Understanding changes in ecological factors, such as decomposition patterns, nutrient cycling, primary productivity, and biodiversity evaluation, depends on changes in the insect community. High-throughput DNA barcoding methods combined with community databases may reduce the number of methodological, budgetary, and logistical obstacles to large-scale examination. Scientists have investigated the Arctic arthropod community, concentrating on Arachnida, Collembola, and Insecta (Pentinsaari et al., 2020).

In one of the case studies, three species from the family of Canidae (one of the most fascinating groups within mammals) were analyzed using the DNA barcoding approach to evaluate genetic differentiation and authenticate COI barcodes. In contrast to the previous studies, the average divergence in the sequence was 12.32% between species, while it was 0.61% within species, indicating a comparatively higher genetic variation (Aksöyek et al., 2017).

Identification of species is very crucial for the protection of animal and plant biodiversity. Ran et al. (2020) conducted a research study using the COI gene as a promising mtDNA barcode to ensure the management of fishery resources as sustainable natural resources. As a result of this research study, 120 species were identified, including three newly identified records in China. These 120 species were distributed to 35 families and seven orders. A comprehensive repository of barcode references was established in GenBank consisting of 306 genetic sequences (GenBank accession numbers MN388943 to MN389209). The results indicated 0.9% intraspecific sequence variation, and 14.7% interspecific variation in genetic sequence.

The reference database of DNA barcodes may help ornithologists improve geographic-based sampling for the identification of COI sequences. Blood samples from 18 different species of birds across 10 groups were analyzed, and 26 COI sequences were generated, which helps the ornithologist reclassify a boreal migratory bird as a discrete species (Pulgarín-R et al., 2021).

Feitosa et al. (2018) conducted a study to track the illegal trade of endangered shark species along Brazil's northern coastline. They analyzed 427 samples using COI (260 samples) and NADH2 (167 samples), identifying 17 distinct species. Among these, they classified 53% as facing extinction, a percentage that would have risen to 76% if near-threatened species had been included (Feitosa et al., 2018).

7.8 CHALLENGES AND LIMITATIONS OF DNA BARCODING

The future of DNA barcoding holds great promise due to advancements in sequencing technologies, increased taxonomic coverage, and its growing role in conservation and policy-making. High-throughput, portable, and long-read sequencing will make DNA barcoding more efficient and accessible. Studying underserved taxonomic groups, including microbes and extinct species, will enhance our understanding of biodiversity. However, challenges remain, such as filling gaps in reference databases in biodiverse regions, and navigating ethical and legal issues related to specimen collection and genetic data. The integration of machine learning and bioinformatics, though promising, needs further refinement for reliable species identification (Odah, 2024)

DNA-based species classification relies on distinguishing genetic variation within species (intraspecific) from that between species (interspecific), though these variation thresholds remain uncertain and may differ across taxa. The absence of a universally conserved yet sufficiently divergent gene for DNA barcoding complicates the identification of recently diverged or hybrid species, emphasizing the need for reference sequences derived from taxonomically verified specimens through collaborative scientific efforts. While concerns persist that DNA barcoding could undermine morphology-based taxonomy and lead to misidentifications solely based on genetic differences, its expanding taxonomic coverage offers immense potential for advancing taxonomy, systematics, and species diagnostics (Schindel & Miller, 2005; Chase et al., 2007).

7.9 FUTURE DIRECTION AND EMERGING TECHNOLOGIES IN DNA BARCODING

DNA barcoding depends on traditional taxonomy to validate species identification and should complement rather than replace conventional taxonomic methodologies. Ensuring fair and widespread access to reference databases and scientific knowledge is vital for enabling underprivileged and developing nations to keep pace with contemporary advancements. While next-generation sequencing (NGS) enhances species differentiation through whole-genome analysis, its primary focus is deciphering genomic intricacy, variation, and functionality rather than facilitating species classification and ecological assessment. DNA barcoding continues to be a cost-efficient, robust, and high-throughput approach for species identification.

There should be more focus on selecting suitable methods specific to species instead of focusing only on the discrimination rates of species. The quality of the reference database can be improved by machine learning because it will reduce the errors resulting from the mislabeling of barcodes. High quality of DNA barcodes may be ensured by an increase in funding for herbaria, which is critical for maintaining vouchers of specimens.

REFERENCES

Ahmed, S. S. (2022). DNA barcoding in plants and animals: a critical review.

Aksöyek, E., İbiş, O., Özcan, S., Moradi, M., & Tez, C. (2017). DNA barcoding of three species (Canis aureus, Canis lupus and Vulpes vulpes) of Canidae. *Mitochondrial DNA Part A*, *28*(5), 747–755.

Antil, S., Abraham, J. S., Sripoorna, S., Maurya, S., Dagar, J., Makhija, S., Bhagat, P., Gupta, R., Sood, U., & Lal, R. (2023). DNA barcoding, an effective tool for species identification: a review. *Molecular Biology Reports*, *50*(1), 761–775.

Bale, J. S., Masters, G. J., Hodkinson, I. D., Awmack, C., Bezemer, T. M., Brown, V. K., Butterfield, J., Buse, A., Coulson, J. C., & Farrar, J. (2002). Herbivory in global climate change research: direct effects of rising temperature on insect herbivores. *Global Change Biology*, *8*(1), 1–16.

Becker, S., Hanner, R., & Steinke, D. (2011). Five years of FISH-BOL: brief status report. *Mitochondrial DNA*, *22*(sup1), 3–9.

Chase, M. W., Cowan, R. S., Hollingsworth, P. M., Van Den Berg, C., Madriñán, S., Petersen, G., Seberg, O., Jørgsensen, T., Cameron, K. M., & Carine, M. (2007). A proposal for a standardised protocol to barcode all land plants. *Taxon*, *56*(2), 295–299.

Chase, M. W., Salamin, N., Wilkinson, M., Dunwell, J. M., Kesanakurthi, R. P., Haidar, N., & Savolainen, V. (2005). Land plants and DNA barcodes: short-term and long-term goals. *Philosophical Transactions of the Royal Society B: Biological Sciences*, *360*(1462), 1889–1895.

Chen, Y., Zhu, X., Loukopoulos, P., Weston, L. A., Albrecht, D. E., & Quinn, J. C. (2021). Genotypic identification of Panicum spp. in New South Wales, Australia using DNA barcoding. *Scientific Reports*, *11*(1), 16055.

Clark, K., Karsch-Mizrachi, I., Lipman, D. J., Ostell, J., & Sayers, E. W. (2016). GenBank. *Nucleic Acids Research*, *44*(D1), D67–D72.

Collins, R., & Cruickshank, R. (2013). The seven deadly sins of DNA barcoding. *Molecular Ecology Resources*, *13*(6), 969–975.

Creer, S., Fonseca, V., Porazinska, D., Giblin-Davis, R., Sung, W., Power, D., Packer, M., Carvalho, G., Blaxter, M., & Lambshead, P. (2010). Ultrasequencing of the meiofaunal biosphere: practice, pitfalls and promises. *Molecular Ecology*, *19*, 4–20.

DeSalle, R. (2007). Phenetic and DNA taxonomy; a comment on Waugh. *Bioessays*, *29*(12), 1289–1290.

Edgar, R. C. (2004). MUSCLE: multiple sequence alignment with high accuracy and high throughput. *Nucleic Acids Research*, *32*(5), 1792–1797.

Elías-Gutiérrez, M., & Valdez-Moreno, M. (2023). Relevance of DNA barcodes for biomonitoring of freshwater animals. *Frontiers in Environmental Science*, *11*, 1057653.

Erickson, D. L., Spouge, J., Resch, A., Weigt, L. A., & Kress, J. W. (2008). DNA barcoding in land plants: developing standards to quantify and maximize success. *Taxon*, *57*(4), 1304–1316.

Feitosa, L. M., Martins, A. P. B., Giarrizzo, T., Macedo, W., Monteiro, I. L., Gemaque, R., Nunes, J. L. S., Gomes, F., Schneider, H., & Sampaio, I. (2018). DNA-based identification reveals illegal trade of threatened shark species in a global elasmobranch conservation hotspot. *Scientific Reports*, *8*(1), 3347.

Galimberti, A., De Mattia, F., Losa, A., Bruni, I., Federici, S., Casiraghi, M., Martellos, S., & Labra, M. (2013). DNA barcoding as a new tool for food traceability. *Food Research International*, *50*(1), 55–63.

Gostel, M. R., & Kress, W. J. (2022). The expanding role of DNA barcodes: indispensable tools for ecology, evolution, and conservation. *Diversity*, *14*(3), 213.

Hebert, P. D., Cywinska, A., Ball, S. L., & DeWaard, J. R. (2003). Biological identifications through DNA barcodes. *Proceedings of the Royal Society of London. Series B: Biological Sciences*, *270*(1512), 313–321.

Hollingsworth, P. M., Li, D.-Z., van der Bank, M., & Twyford, A. D. (2016). Telling plant species apart with DNA: from barcodes to genomes. *Philosophical Transactions of the Royal Society B: Biological Sciences*, *371*(1702), 20150338.

Imtiaz, A., Nor, S. A. M., & Naim, D. M. (2017). Progress and potential of DNA barcoding for species identification of fish species. *Biodiversitas Journal of Biological Diversity*, *18*(4), 1394–1405.

Janda, J. M., & Abbott, S. L. (2007). 16S rRNA gene sequencing for bacterial identification in the diagnostic laboratory: pluses, perils, and pitfalls. *Journal of Clinical Microbiology*, *45*(9), 2761–2764.

Kaur, S. (2015). DNA barcoding and its applications. *International Journal of Engineering Research and General Science*, *3*(2), 602–604.

Kesanakurti, P. R., Fazekas, A. J., Burgess, K. S., Percy, D. M., Newmaster, S. G., Graham, S. W., Barrett, S. C., Hajibabaei, M., & Husband, B. C. (2011). Spatial patterns of plant diversity below-ground as revealed by DNA barcoding. *Molecular Ecology*, *20*(6), 1289–1302.

Kõljalg, U., Larsson, K. H., Abarenkov, K., Nilsson, R. H., Alexander, I. J., Eberhardt, U., Erland, S., Høiland, K., Kjøller, R., & Larsson, E. (2005). UNITE: a database providing web-based methods for the molecular identification of ectomycorrhizal fungi. *New Phytologist*, *166*(3), 1063–1068.

Kress, W. J., & Erickson, D. L. (2012). *DNA barcodes: methods and protocols*. Springer.

Kress, W. J., Wurdack, K. J., Zimmer, E. A., Weigt, L. A., & Janzen, D. H. (2005). Use of DNA barcodes to identify flowering plants. *Proceedings of the National Academy of Sciences*, *102*(23), 8369–8374.

Krishna Krishnamurthy, P., & Francis, R. A. (2012). A critical review on the utility of DNA barcoding in biodiversity conservation. *Biodiversity and Conservation*, *21*, 1901–1919.

Kumar, S., Stecher, G., Peterson, D., & Tamura, K. (2012). MEGA-CC: computing core of molecular evolutionary genetics analysis program for automated and iterative data analysis. *Bioinformatics*, *28*(20), 2685–2686.

Lim, B. T., Sadanandan, K. R., Dingle, C., Leung, Y. Y., Prawiradilaga, D. M., Irham, M., Ashari, H., Lee, J. G., & Rheindt, F. E. (2019). Molecular evidence suggests radical revision of species limits in the great speciator white-eye genus Zosterops. *Journal of Ornithology*, *160*, 1–16.

Luo, A., Zhang, A., Ho, S. Y., Xu, W., Zhang, Y., Shi, W., Cameron, S. L., & Zhu, C. (2011). Potential efficacy of mitochondrial genes for animal DNA barcoding: a case study using eutherian mammals. *BMC Genomics*, *12*, 1–13.

Mcleod, D. S., Horner, S. J., Husted, C., Barley, A., & Iskandar, D. (2011). "Same-same, but different": an unusual new species of the Limnonectes kuhlii complex from West Sumatra (Anura: Dicroglossidae). *Zootaxa*, *2883*(1), 52–64.

Mir, R. A., Bhat, K. A., Rashid, G., Ebinezer, L. B., Masi, A., Rakwal, R., Shah, A., & Zargar, S. M. (2021). DNA barcoding: a way forward to obtain deep insights about the realistic diversity of living organisms. *The Nucleus*, *64*, 157–165.

Mishra, P., Kumar, A., Nagireddy, A., Mani, D. N., Shukla, A. K., Tiwari, R., & Sundaresan, V. (2016). DNA barcoding: an efficient tool to overcome authentication challenges in the herbal market. *Plant Biotechnology Journal*, *14*(1), 8–21.

Murphy, R. W., Crawford, A. J., Bauer, A. M., Che, J., Donnellan, S. C., Fritz, U., Haddad, C. F., Nagy, Z. T., Poyarkov, N. A., & Vences, M. (2013). *Cold code: the global initiative to DNA barcode amphibians and nonavian reptiles*. Wiley Online Library.

Nakahama, N., Isagi, Y., & Ito, M. (2019). Methods for retaining well-preserved DNA with dried specimens of insects. *European Journal of Entomology*, *116*, 486–491.

Odah, M. (2024). Advancements in DNA barcoding: revolutionizing taxonomy and biodiversity studies.

Pang, X., Liu, C., Shi, L., Liu, R., Liang, D., Li, H., Cherny, S.S. and Chen, S., 2012. Utility of the trnH–psbA intergenic spacer region and its combinations as plant DNA barcodes: a meta-analysis. *PloS one*, *7*(11), p.e48833.

Parmentier, I., Duminil, J. M., Kuzmina, M., Philippe, M., Thomas, D. W., Kenfack, D., Chuyong, G. B., Cruaud, C., & Hardy, O. J. (2013). How effective are DNA barcodes in the identification of African rainforest trees? *PLoS One*, *8*(4), e54921.

Pawlowski, J., Audic, S., Adl, S., Bass, D., Belbahri, L., Berney, C., Bowser, S. S., Cepicka, I., Decelle, J., & Dunthorn, M. (2012). CBOL protist working group: barcoding eukaryotic richness beyond the animal, plant, and fungal kingdoms. *PLoS Biology*, *10*(11), e1001419.

Pentinsaari, M., Blagoev, G. A., Hogg, I. D., Levesque-Beaudin, V., Perez, K., Sobel, C. N., Vandenbrink, B., & Borisenko, A. (2020). A DNA barcoding survey of an Arctic arthropod community: implications for future monitoring. *Insects*, *11*(1), 46.

Piper, A. M., Batovska, J., Cogan, N. O., Weiss, J., Cunningham, J. P., Rodoni, B. C., & Blacket, M. J. (2019). Prospects and challenges of implementing DNA metabarcoding for high-throughput insect surveillance. *GigaScience*, *8*(8), giz092.

Porter, T. M., & Hajibabaei, M. (2018). Over 2.5 million COI sequences in GenBank and growing. *PLoS One*, *13*(9), e0200177.

Pulgarín-R, P. C., Olivera-Angel, M., Ortíz, L., Nanclares, D., Velásquez-Restrepo, S., & Díaz-Nieto, J. F. (2021). DNA barcodes of birds from Northern Colombia. *Biodiversity Data Journal*, *9*.

Pyšek, P., Hulme, P. E., Meyerson, L. A., Smith, G. F., Boatwright, J. S., Crouch, N. R., Figueiredo, E., Foxcroft, L. C., Jarošík, V., & Richardson, D. M. (2013). Hitting the right target: taxonomic challenges for, and of, plant invasions. *AoB Plants*, *5*.

Ran, K., Li, Q., Qi, L., Li, W., & Kong, L. (2020). DNA barcoding for identification of marine gastropod species from Hainan Island, China. *Fisheries Research*, *225*, 105504.

Ratnasingham, S., & Hebert, P. D. (2007). BOLD: the barcode of life data system (www.barcodinglife.org). *Molecular Ecology Notes*, *7*(3), 355–364.

Ratnasingham, S., & Hebert, P. D. (2013). A DNA-based registry for all animal species: the barcode index number (BIN) system. *PLoS One*, *8*(7), e66213.

Sayers, E. W., Cavanaugh, M., Clark, K., Ostell, J., Pruitt, K. D., & Karsch-Mizrachi, I. (2020). GenBank. *Nucleic Acids Research*, *48*(D1), D84.

Schindel, D. E., & Miller, S. E. (2005). DNA barcoding a useful tool for taxonomists. *Nature*, *435*(7038), 17–17.

Shen, Y.-Y., Chen, X., & Murphy, R. W. (2013). Assessing DNA barcoding as a tool for species identification and data quality control. *PLoS One*, *8*(2), e57125.

Simberloff, D., Martin, J.-L., Genovesi, P., Maris, V., Wardle, D. A., Aronson, J., Courchamp, F., Galil, B., García-Berthou, E., & Pascal, M. (2013). Impacts of biological invasions: what's what and the way forward. *Trends in Ecology & Evolution*, *28*(1), 58–66.

Tamura, K., Peterson, D., Peterson, N., Stecher, G., Nei, M., & Kumar, S. (2011). MEGA5: molecular evolutionary genetics analysis using maximum likelihood, evolutionary distance, and maximum parsimony methods. *Molecular Biology and Evolution*, *28*(10), 2731–2739.

Teodoro, S., Terossi, M., Mantelatto, F., & Costa, R. (2016). Discordance in the identification of juvenile pink shrimp (Farfantepenaeus brasiliensis and F. paulensis: Family Penaeidae): an integrative approach using morphology, morphometry and barcoding. *Fisheries Research*, *183*, 244–253.

von Cräutlein, M., Korpelainen, H., Pietiläinen, M., & Rikkinen, J. (2011). DNA barcoding: a tool for improved taxon identification and detection of species diversity. *Biodiversity and Conservation*, *20*, 373–389.

Wang, T., Zhang, Y.-P., Yang, Z.-Y., Liu, Z., & Du, Y.-Y. (2020). DNA barcoding reveals cryptic diversity in the underestimated genus Triplophysa (Cypriniformes: Cobitidae, Nemacheilinae) from the northeastern Qinghai-Tibet Plateau. *BMC Evolutionary Biology*, *20*, 1–15.

Waugh, J. (2007). DNA barcoding in animal species: progress, potential and pitfalls. *BioEssays*, *29*(2), 188–197.

Will, K. W., Mishler, B. D., & Wheeler, Q. D. (2005). The perils of DNA barcoding and the need for integrative taxonomy. *Systematic Biology*, *54*(5), 844–851.

Williams, J., McKay, S., Khalfan, M., Hilgert, U., Lauter, S., & Jeong, E. (2014). *DNA subway–an educational bioinformatics platform for gene and genome analysis: DNA barcoding, and RNA-Seq*. Proceedings of the 10th World Congress on Genetics Applied to Livestock Production, Vancouver, BC, Canada.

Yang, F., Ding, F., Chen, H., He, M., Zhu, S., Ma, X., Jiang, L., & Li, H. (2018). DNA barcoding for the identification and authentication of animal species in traditional medicine. *Evidence-Based Complementary and Alternative Medicine*, *2018*.

Yu, J., Wu, X., Liu, C., Newmaster, S., Ragupathy, S., & Kress, W. J. (2021). Progress in the use of DNA barcodes in the identification and classification of medicinal plants. *Ecotoxicology and Environmental Safety*, *208*, 111691.

Zhu, S., Liu, Q., Qiu, S., Dai, J., & Gao, X. (2022). DNA barcoding: an efficient technology to authenticate plant species of traditional Chinese medicine and recent advances. *Chinese Medicine*, *17*(1), 112.

8 Prospects of DNA Barcoding in Protists for Exploring and Conserving Their Biodiversity

Imran Ul Haq, Siddra Ijaz, Muhammad Asif Saddique, Ayesha Ashfaq, and Abdullah

8.1 INTRODUCTION

DNA barcoding is a technique that uses a short, standardized portion of the genome to identify and classify species. It works similarly to a barcode scanner in a supermarket, which reads the unique pattern of bars in a barcode to determine what type of product it belongs to (Mathur et al., 2021). In this case, however, we use a specific DNA sequence from a relatively short portion of the genome to distinguish and classify species. The routinely used region for animals is the mitochondrial cytochrome c oxidase I gene, while for plants, it is often a portion from the chloroplast genome, such as *rbcL* and *matK*. Finally, this region is usually the internal transcribed spacer region of ribosomal DNA for fungi. These regions were chosen because they are relatively constant within a species and vary between them (Lahaye et al., 2008).

Historical Context: Paul Hebert and his colleagues first proposed DNA barcoding. In their paper published in 2003, they presented the concept that a single gene region can be used as a universal barcode of animal species and enable very fast and precise identification of a species (Hebert et al., 2003). It was a very innovative notion because, for many years, theories and experiments in molecular biology and genetics began to prove the possibility and expediency of using DNA sequencing as a tool for understanding biodiversity. As such, before Hebert's research, species identification was based on morphological features and depended on taxonomists' ability. Although it was very efficient, this method was very slow and required a lot of expertise, which was not always sufficient due to the existence of cryptic species and the great phenotypic plasticity of organisms. Sequencing technologies in the late 20th century allowed researchers to overcome these constraints (Krehenwinkel et al., 2019).

DOI: 10.1201/9781003538929-8

In the early 2000s, there was growing interest in discovering a short DNA sequence that would be a universal marker for species identification. As a genome unit, mitochondria became a useful region for such a task, although usual criticism was related to prokaryotes with no mitochondria. Hebert's research on COI in animals proved that it was a very effective gene because there was an exceptional amount of divergence in the gene's regions among closely related species. They also found that this method can work in other taxonomic groups. As such, there was a need to create infrastructure to keep the barcode sequence and organism information. That is why the Barcode of Life Data System was launched in 2005 and is now widely used. Since the concept of DNA barcoding emerged, it has been applied to a wide range of species, including animals, plants, fungi, and microorganisms. It became an indispensable tool in taxonomy, ecology, conservation biology, and many other disciplines (Hebert and Gregory, 2005). The International Barcode of Life project, launched in 2010, aims to sequence millions of species and create a comprehensive library with reference samples available (Ratnasingham and Hebert, 2013).

8.1.1 Implications of DNA Barcoding

One of the greatest advantages of DNA barcoding is that the method helps simplify and improve species identification. Traditional taxonomic methods require personal judgment and research to be completed. Many species in certain areas are not studied confidently, and the identification of species in some areas can take years and decades. Moreover, some species remain unidentified because of the lack of understanding among the taxonomists. DNA barcoding defies traditional taxonomy's subjective difficulties by giving an objective and efficient method of species identification. DNA barcoding is ideal for regions with abundant biodiversity and scarce taxonomic knowledge of this biodiversity (Barcaccia et al., 2016).

DNA barcoding is especially helpful in identifying cryptic species and life stages, so these species can often be separated only by DNA barcoding. Identification is impossible in the case of insect larvae, while barcoding may help distinguish the larvae from their corresponding adult forms. Similarly, in the case of protists and other microorganisms, it is possible to identify different species with the help of DNA barcoding because these species do not have meaningful formalized characteristics to be segmented according to them (Mulvaney et al., 2023).

Exploration of Evo-Eco dynamics: It provides deep insights into species' evolutionary and ecological existence. Firstly, by analyzing the barcode sequence, scientists can determine the phylogenetic relationship and establish a better tree of life that reflects a more accurate representation of molecules. Consequently, it enhances understanding of species' evolutionary history and divergence (Hebert et al., 2003). Moreover, DNA barcoding is adopted in ecology. Complexity studies related to genetic information were used to discover various food webs and species interactions. More specifically, it can be used to analyze gut content and find the eating habits of herbivores, predators, and species. Thus, countless species interactions can be detected, including trophic relationships and patterns of energy flow (Yu et al., 2023).

Moreover, it is vital to admit that DNA barcoding is not limited to taxonomy and biodiversity work. One of the most common applications is agriculture. Establishing proper and sustainable pest management measures requires identifying and detecting pests and natural enemy species. Mistakes in identification are fatal, as they result in inefficient conventional pest management, and ultimately increase pesticide use (Jalali and Ojha, 2015). Similarly, identification is suitable, and coatings can also be used to guarantee food safety. The medical field has also made advances in DNA barcoding. For instance, knowing the disease vector helps identify many mosquito species, including malaria and dengue fever. These measures have an unfathomable value number and prevent certain diseases (Dawan and Ahn, 2022).

Though DNA barcoding has many advantages, various obstacles exist. One of the most significant impediments is the issue of selecting the right barcode regions. Choosing suitable markers remains a debate for certain groups of organisms, and it is not always possible to identify the best sequencing primers. Besides, it is also important to note that some species may be too similar for their sequences to be resolved by DNA barcoding. In such cases, other genetic markers or WGS can be used. The good quality of the reference databases also plays an important role in the absence of certain haplotypes in the barcoding databases, which prevents their reliable recognition. It is especially important in the case of poorly studied groups, as the absence of a reference dataset can mean that the barcoding analysis would not be helpful (Cheng et al., 2023).

In addition, the effectiveness of DNA barcoding depends on the advancement and precision of basic sciences such as molecular biology and genetics. These scientific fields will continue to develop in the future, so there is no doubt that the difficulties of applying the technology will decrease. It is important to note that DNA barcoding has great potential because of the development of sequencing technologies and bioinformatics. In particular, next-generation sequencing allows the simultaneous determination of the sequences of several, or even dozens of barcode regions, or entire genomes. The data obtained can then be analyzed with various computational tools, including machine learning. The future development of these techniques should guarantee that DNA barcoding will be widely used in biodiversity science and beyond (Odah, 2023).

8.2 DNA BARCODING

Concept of DNA Barcoding: DNA barcoding is a modern approach to identifying various species by analyzing a particular genome fragment. The principal positive aspect of DNA barcoding is that it proposes a strategy that could be universal, uniform, and versatile to cover different organisms and to be applied in various conditions. DNA barcoding primarily aims to find fragments that would vary significantly between species while being relatively stable within the species. The fragment can then be called a 'barcode' for this species. However, the region to be analyzed should meet specific requirements, such as the ability to be easily sequenced and amplified by conventional laboratory technologies, and to vary to allow defining the difference between closely related species (Cowan and Fay, 2012).

Approaches to DNA Barcoding: The methodology of DNA barcoding consists of different critical steps: sample collection and preservation, DNA extraction, PCR amplification, sequencing, and data analysis. Each step is vital to identifying species and must be followed with meticulous attention to detail to ensure the accuracy and reliability of the results (Stein et al., 2013).

1. ***Sample Collection and Preservation:*** To begin the process, various biological samples must be collected from the target organisms. These can represent tissues, cells, or even environmental samples containing DNA traces. However, to use these samples or to store them effectively, it is necessary to focus on proper preservation: all samples need attention to preservation details to prevent DNA loss, damage, or contamination.
2. ***DNA Extraction:*** The next step involves using various chemical and mechanical methods to extract the DNA from the collected samples. The purpose is to obtain relatively pure and undamaged DNA that can be used in the following processes.
3. ***PCR Amplification:*** PCR amplification is used to amplify the extracted DNA. The goal is to carefully duplicate a desirable DNA region to produce enough amounts for sequencing.
4. ***Sequencing:*** The next step is sequencing the amplified DNA to determine the exact sequence of the nucleotides in the barcoding region or barcode locus. Currently, advances in sequencing technology are notable, making the process faster, more efficient, and cost-effective.
5. **Data Analysis:** Finally, the collected sequences are analyzed, and the species is identified by exploring the reference sequences in special databases. At this point, a range of bioinformatics tools can be used for sequence alignment, calculation of genetic distances, and development of phylogenetic trees (Kress and Erickson, 2012).

8.2.1 DNA Barcode Markers

The literature describes different markers used for DNA barcoding that can correctly differentiate a particular species from all others. Each group requires specific markers since various organisms have different genetic compositions and development.

Cytochrome c oxidase I: COI is one of the most known animal markers and part of a mitochondrial genome. It was first proposed for use by Paul Hebert et al. in 2003. The COI gene is about 650 base pairs long; it synthesizes a polypeptide belonging to the essential catalytic subunit of the enzyme cytochrome c oxidase, cattle, gr. The latter is vital for the cellular respiration process, relaying electrons. This gene was observed to have a very large variation between species and relatively small variation within the species. Its maternal inheritance and the absence of the recombination process between mother and father polymorphisms make it an essential marker for defining a species.

Internal transcribed spacer (ITS): The ITS region is the most known and widely used barcode marker for fungi between small and large subunits of ribosomal RNA genes. It has high variability, which increases the accuracy of species determinacy. In addition, it has a conserved region of the 5.8S ribosomal RNA gene and variable areas of ITS1 and ITS2 to meet these purposes. As for additional indicators, sometimes the second loci of the gene are used to barcode more accurately. For example, during bacterial and archaea DNA barcoding, 16S rRNA is used, while for protists, it is 18S rRNA. The researchers selected these genes for thousands of years as the most informative about the difference in species. In addition, the other researchers could compare the results using the same samples (Nurhasanah, 2019).

Ribulose-1,5-bisphosphate carboxylase oxygenase large subunit (rbc L) and Maturase K (mat K): The chloroplast genes make plants' barcoding markers. One of them, *rbcL*, a composition of about 1400 bp, synthesizes a bigger subunit of the rubisco enzyme. The latter is important in Calvin cycle plants, so ribulose 1,5-bisphosphate. It is a valuable distinguishing tool. The higher variation of *matK* gene product, about 800 bp long, is in no way less; it synthesizes essential protein for group II introns' splicing, making the plant samples' definition possible. The evaluation used two particular genes: plastid *rbcL* (ribulose 1,5-bisphosphate carboxylase/oxygenase large subunit) and *matK* (Maturase K).

8.2.2 Protista

Protists are a diverse array of eukaryotic microorganisms that cannot be clearly defined as plants, animals, or fungi in conventional subdivisions. Primarily, protists comprise single-celled organisms, including colonial and unpretentious multicellular forms (Triplett et al., 2023). In other words, the kingdom of Protista is often seen as a blanket term for all eukaryotes that do not belong to other eukaryotic kingdoms. Among the main properties of protists that differentiate them from prokaryotic organisms are their complex cellular structure and the presence of a nucleus and numerous membrane-bound cell organelles. Protists come in an enormous variety of forms and shapes. They can be distinguished based on their morphology and structures. Such forms are amoeboid, flagellated, ciliated, or sessile, with many having impressive shells or protective coverings. Protists can photosynthesize, consume organic matter, or do both. This feature allows them to adjust their nutrition mode and thrive in many ecological niches. Protists can reproduce sexually and asexually. Asexually, they reproduce through binary fission, budding, or spore production. Sexual reproduction is possible by merging gametes and the following recombination of genes. Protists can be found in freshwater, marine, and soil biomes. They come in various forms and substances, ranging from the least salty water to the most concentrated and acidic combinations. Protists can survive in a vast range of habitats and often exist in extreme unlikeliness: protists survive in highly saline, sour, or hot environments (Corliss, 2002).

8.2.2.1 Classification of Protists

Protists are a huge medical group with numerous forms, and concepts of their nature are diverse and various. In this case, scientists employ the most modern and recent molecular findings and reports to update and modify the taxonomy. Traditionally, it is usual to differentiate and categorize protists based on their resemblance to plants, animals, or fungi.

8.2.2.2 Traditional Classification

1. Protozoa, which embody animal-like protists, are mostly seen as direct predators of other living organisms and can move.
2. Algae and other plant-like protists are grass-like and usually perform photosynthesis (Whittaker and Margulis, 1978).

They can be multifaceted and unicellular; examples of such species are seaweed and algae.

8.2.2.3 Modern Classification

Progress in molecular biology and phylogenetics has established a better understanding of protist diversity. The protist classification is often based on genetic association and grouped into different supergroups. These are the largest assemblages for which we can identify clear genetic differences and sequence similarities. Some of the most notable supergroups include:

1. Excavata: This supergroup includes heterotrophs such as *Euglena* and *Trypanosoma*, named for a feeding groove and the corkscrew mobile structure in their flagella.
2. The SAR group comprises Stramenopila, foraminifera, and Alveolata.
3. Archeplastida: Red algae, green algae, and land plants. This supergroup underscores the availability of specific types of algae and land plants.
4. Amoebozoa: These are the protists that move by amoeboid motion; this supergroup comprises slime molds and several other amoebas.
5. Opistoconts: Fungi, animals, and choanoflagellates (Adl et al., 2012).

8.2.3 Ecological Roles and Importance of Protists

Protists have various ecological roles in different ecosystems. These organisms fuel primary production, nutrient cycling, and the control of different microbial communities, as shown in Figure 8.1. The ecological significance of these organisms is discussed below:

i. **Primary Production:** Photosynthetic protists are the primary producers in both terrestrial and aquatic ecosystems. Examples include diatoms, dinoflagellates, and specific green algae species. These organisms fuel the food web. For example, higher organisms such as zooplankton, fish, and other marine organisms depend on photosynthetic protists to survive. They convert sunlight into organic substances that the higher trophic levels can consume. The process fuels primary production, thereby playing a critical role in different ecosystems (Ares, 2019).
ii. **Nutrient Cycling:** Several protists are associated with the decay of organic substances. *Purkalec* and *Neustupa* posit that molds and slime molds of water molds disintegrate dead organic matter, thereby ensuring nutrient recycling. The role of protists in the decomposition process is sufficient to preserve soil fertility and maintain the productivity of different ecosystems (Silva, 2020).

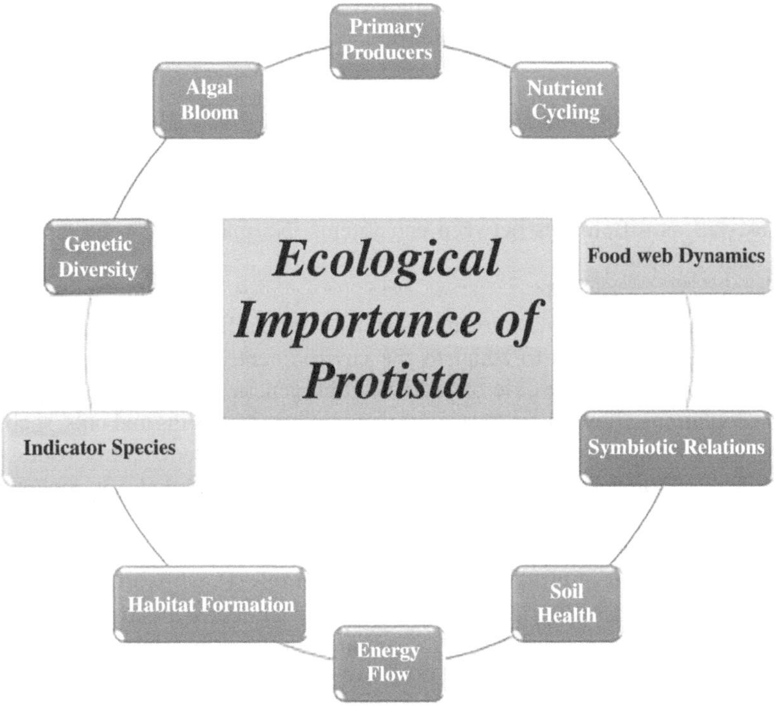

FIGURE 8.1 Ecological importance of Protists: Protists are vital components of ecosystems because they function as nutrient recyclers while playing fundamental roles in ecosystem food webs and ecology. They also maintain healthy soil conditions. Ecosystems receive influence from protists through symbiotic relationships and algal blooming events. Protists serve as biological markers while influencing genome diversity.

iii. **Symbiotic Roles:** It is agreeable that protists form symbiotic relationships with specific organisms. For example, certain types of ciliates and flagellates are normally found in the digestive systems of termites and ruminants. The organisms play a positive role in the breakdown of cellulose. Another example of a symbiotic relationship is the case of corals and photosynthetic dinoflagellates or zooxanthellae (Muñoz et al., 2018).

8.2.3.1 Role in Environmental Indicators

Protists can be used to indicate environmental health and water quality. The shift in the occurrence of protists provides critical information on emerging environmental issues such as pollution, nutrient influxes, and changing climate. Monitoring these organisms is, therefore, useful in determining the dynamics of an ecosystem and the effectiveness of conservation efforts. Some protists often cause diseases in humans, animals, and plants. For instance, *Plasmodium*, *Trypanosoma*, and *Phytophthora* are important kinds that often cause malaria, sleeping sickness, and potato blight, respectively. Understanding these protists is crucial for developing effective treatment and management approaches (Fournier et al., 2022).

8.3 BIODIVERSITY OF PROTISTA

Biodiversity generally refers to the wealth of life on Earth and includes variety and variability. Its components include differences within other species and other species, as well as differences between ecosystems. Normally, it can be divided into three main types:

1. Genetic variation refers to intra-species genetic differences. Such a state enables populations to adapt to the environment, withstand disease, and remain viable in the case of protists. High genetic diversity within protist populations allows them to survive their environment's ins and outs, such as the changing temperature, pH, and salinity level. Moreover, it provides for the development of new features, allowing protists to exploit new ecological niches. Another topic of genetic variation is mutation, recombination, and transfer of genetic material that applies to other new variants, new features, and extinctions of protists within the populations (Hawksworth, 2009).
2. Species Diversity is rich in species and is defined as relatively small or larger areas or environments. Protists handle an immense amount of species diversity, with 200,000–1,000,000 or even more species known. They live in various habitats, from freshwater streams and ponds to extreme ones like hot springs and polar ice. Scientists try to differentiate protists in several ways, such as foodstuffs and appearance. Most known protist species in the world fall into two main groups, which are five groups of photosynthetic algae that produce food for most other forms of life, zooplankton, and other mixotrophic protists that ingest bacteria, algae, and other rinse droplets in the water, and other materials. The number of species shows how vast protists are, where they can be found in most biological habitats and niches (Carvalho da Silva and Fernandes, 2023)
3. Ecosystem Diversity includes different types of ecosystems in space. Protists are responsible for chemical and energy flows in all kinds of ecosystems. They multiply and return nutrients and produce food for animals. They decompose organic matter and make it available for other cells. They also influence other organisms' ages, ultimately affecting many different systems. Ecosystem diversity among the protists can extend to the fact that they can weather a wide range of environments (Firkins et al., 2020).

8.3.1 Genetic Diversity of Protists

Regarding protists, a significant source of diversity is their genetics. It supports their ability to adapt to various environments and to possess distinctive features. Genetic mechanisms can be identified, including mutations, recombination, and horizontal gene transfer. First, mutations are the random changes in DNA sequences, and although the creation of a new gene is rare, it is assumed that some changes can make a protist better adapted to the environment, for example, immune to a type of pathogen (Lebret et al., 2012). The second mechanism of genetic diversity, specifically sexual reproduction, meant that this genetic recombination is beneficial because it fuses both parents' gametes, resulting in their combination. Finally, horizontal gene transfer is the third and most important source of genetic diversity, which can introduce entirely new genes and functions. In such a way, the protists possess notable genetic diversity that can allow for their adaptation and survival in various environments (El-Diasty, 2018).

An example of the diversity of protist species is photosynthetic protists, which represent the primary producers in oceans, converting sunlight into organic matter and supporting food webs. Photosynthetic protists, or algae, include multiple forms with different pigments, for instance, brown diatoms and red dinoflagellates. On the other hand, heterotrophic protists, or protozoans, a population of amoebas and ciliates, feed on bacteria, algae, and other small organisms to preserve ecological homeostasis, control microbial populations, and recycle. Such organisms cannot create organic matter but still greatly impact the Earth's ecosystems (Tate et al., 2019).

8.3.2 Ecosystem Diversity and Roles of Protists

Protists play many roles in an ecosystem's functioning and diversity, with some supporting the productivity of these ecosystems. Primarily, photosynthetic protists support many ecosystems by forming the basis of the food web. For example, marine and freshwater ecosystems can sustain a wide range of consumers, ranging from zooplankton to fish and marine mammals, because of algae (Tate et al., 2019). Photosynthetic protists help in the cycling of carbon through carbon dioxide sequestration and when they die and settle at the bottom of the oceans. The accumulation of their sediments then leads to the formation of carbon-rich sediments in the marine bottom. In terrestrial ecosystems, protists such as chlorophytes are essential for the cycling of nutrients. Absorbed and consumed by predators and other organisms, they help incorporate nutrients into their bodies, making them available to plants when they die and decompose (Firkins et al., 2020). On the other hand, in aquatic and terrestrial ecosystems, there are protists other than algae but amoebas and flagellates, which are primary consumers of bacteria (Chandarana and Amaresan, 2022).

Protists' genetic, species, and ecosystem biodiversity are relevant to their varying degrees of adaptability, resilience, and similar properties, helping them and contributing to their ecological significance. Genetic diversity allows protists to adapt to various climates and develop new characteristics, promoting their success in each new environment. The diversity of species reflects on the organisms' ability to occupy different niches in the natural world and perform multiple ecological functions necessary for the ecosystems' integrity and proper function. Protists are crucial agents participating in the processes of primary production, the cycling of nutrients, and the mutualism and symbiosis with other organisms. They are all related to building and maintaining structures and functions of the ecosystems. Additionally, they are needed to ration and formulate laws concerning other ecosystem members. Therefore, saving is the only way to protect other organisms from being endangered (Carvalho da Silva and Fernandes, 2023).

8.4 APPROACHES TO INVESTIGATING PROTIST BIODIVERSITY

Studying protist diversity is vital for comprehending their ecological roles, evolutionary patterns, and opportunities for their application. However, due to their small size and numerous types, the variety of protists is a significant challenge to study for biodiversities (Burki et al., 2021). Consequently, to tackle the task, researchers utilize various methods, including traditional ones, such as microscopy and culturing, and more advanced options, such as DNA barcoding, metagenomics, and transcriptomics.

8.4.1 Conventional Techniques

Certainly, such traditional techniques as microscopy and culturing have always laid the foundation for studying protists' biodiversity. Microscopy allows protists to be examined directly through light and electron microscopes, exploring their bodies' physical features. While light microscopy is applicable and allows viewing the protists, it is limited by the resolution, rendering the examination of the highly complex structure of smaller protists impossible. At the same time, electron microscopy provides high resolution and offers a detailed examination of the cell structures. However, it has serious complications in the sample preparation process and does not work with live samples. Therefore, microscopy is still heavily employed for identifying and taxonomically classifying protists, providing valuable information on their biodiversity and ecological roles (Weydmann-Zwolicka et al., 2024).

Culturing is a traditional technique based on isolating protists and growing them in laboratory conditions to examine their biology and interactions. Thus, culturing allows us to isolate specific protist species in mixed environmental samples and conduct thorough studies of their physiology, reproduction patterns, and responses to circumstances. Cultures can be maintained for significant periods, allowing for multiple investigations and experiments. However, the process is selective, as some species are more easily cultivated than others. Furthermore, laboratory conditions differ from the environment, which might affect protist behavior and physiology. Nevertheless, the process is still highly valuable for protist biodiversity studies, as it facilitates experiment conduction and long-term studies (Ahmed, 2014).

8.4.2 Modern Molecular Techniques

Recent advances in molecular biology have changed the protist biodiversity research field. They lead to introducing more convenient, less expensive, and more accurate methods to identify protists and learn more about them. This molecular technique uses a short, well-established region of DNA that can be employed to determine the exact species of organism. Protists use different regions to barcode the species, the most popular being the internal transcribed spacer region, the small subunit ribosomal RNA, and cytochrome c oxidase I genes (Hoef-Emden, 2012). DNA barcoding is the most accurate method to identify the species, completely independent of their external features. This technique is also useful for uncovering cryptic species that look identical but are completely different genetically. It is also a useful tool because it can automate and expand biodiversity (Smith et al., 2005). On the negative side, however, the identification depends mostly on precise and extensive reference databases, resulting in a dead end if considered inadequate. In addition, various technical problems may also be a challenge, from the impossibility of amplifying and sequencing the barcode regions to those that are now unknown or impossible to overcome (Elsaied et al., 2021).

Metagenomics is the culture-independent method to sequence combined genomes of all species from environmental samples. Although this technology is mostly used to characterize bacteria, it is also a valuable tool in protist research. In fact, with a metagenomics tool, many responses have been given to questions that were impossible or very difficult to answer. Many researchers have focused on the wide array of protists that cannot be cultured. The new genetic methods have shed some light on their modes of life and roles in nature by analyzing the subtypes and functions of the newly found species. Even though the tool of metagenomics is very convenient, analyzing the data sequenced constitutes a very difficult task, requiring advanced computational technologies and essential bioinformatics experience. The quality of the DNA received also depends on its extraction method. Finally, modern high-throughput sequencing, or next-generation sequencing, is expensive, making it impossible to sequence all the diversity of environmental samples (Dawson and Fritz-Laylin, 2009).

Transcriptomics is the approach to analyzing RNA transcripts, and it helps researchers to study gene expression patterns. For protists, it is a useful tool that can add more information about their functional biology and provide an understanding of their adaptation to different conditions. This type of analysis shows the genes characterized by high-level activity. As a result, it helps to understand the potential and its dynamics in the case of comparison. In addition, transcriptomics allows one to consider the changes in the expression of one or several genes during a specific period. In the context of protists, this time may be associated with stress conditions. Finally, comparing the characteristics of different species leads to identifying the common and unique features or genes for specific conditions and species (Adams, 2008).

As for the disadvantages, it is important to note that RNA is less stable than DNA, and it can be more difficult to maintain samples and extract nucleic acids from them. At the same time, deeply sequenced transcriptomes imply the existence of low-abundance transcripts. Thus, the sequencing process may require additional expenses and be more complex (Chheda et al., 2024). Finally, transcriptomic analysis is based on bioinformatics approaches that, along with the necessity of particular software and hardware, are also associated with the requirement of the relevant level of knowledge and skills. In conclusion, the study of protist biodiversity uses traditional and modern methods frequently associated with various benefits and drawbacks. Older approaches like microscopy and culturing allow one to learn more about their morphology and biology. Modern molecular techniques like DNA barcoding, metagenomics, and transcriptomics help researchers identify and consider the characteristics of most protists and can be associated with a more reliable and profound understanding of protists and biodiversity in general (Deshpande, 2023).

8.5 APPLICATIONS OF DNA BARCODING IN PROTISTA

DNA barcoding is an essential technique in protistology. This method has advanced the identification, classification, and understanding of protists, diverse, and ecologically important microorganisms.

The most important aspect of the relevance for protists is that DNA barcoding can provide a precise, unbiased, and rapid tool for species identification. The prior overlooking of the scattered scientific literature poses a particular issue, as various protist taxa have been described based on variable and often archaic features besides the firmness and the hidden nature of many species. DNA barcoding, nonetheless, apprehended these issues using the genetic information consistent within a given species and varies clearly between species. For protists, the development of this modern method was particularly beneficial for identifying look-alike species and those which, perhaps, uninfected or in other life stages had to change their appearance. The discovery of DNA barcoding mostly transformed the study of the diversity of all scholars outside protistology, biologists, and ecologists, since it has enabled the identification of species that are both rapid and accurate, even for complex samples, wherein numerous species coexist or live in the same environment (Gamit and Amin, 2024).

Additionally, DNA barcoding aids in the creation of comprehensive reference databases, such as the Barcode of Life Data System, which is a compilation of barcodes attached to taxonomic information. These databases are critical for identifying species and comparing their sequences across studies and regions. Thus, researchers can utilize and contribute to these databases to enhance the precision and consistency of protist identification, which may be relevant for biodiversity inventories, ecological assessments, and conservation activities. Moreover, the applications of DNA barcoding are not limited to taxonomy, and this technique could be beneficial for various practical needs. Some include identifying invasive species, monitoring ecosystems' health, and evaluating environmental changes' effects on protist biota (Hebert and Gregory, 2005).

However, DNA barcoding for protists has some disadvantages despite numerous benefits. One of the significant challenges arises from the high genetic variability of these populations. The many options available for barcoding make it challenging to choose the best barcode regions and measure genetic distances between sequences. In addition, there might be some cases where the commonly used barcodes, such as the internal transcribed spacer and the small subunit ribosomal RNA gene, may lack sufficient resolution to differentiate between closely related species. Thus, it is essential for researchers to carefully choose and validate the barcode regions for different groups of protists to ensure accurate species identification (Pawlowski et al., 2012).

Another challenge is that many protists have complex life cycles with numerous visually distinct stages. Varying life stages must be linked to the same species using DNA barcoding for their genetic differences in many species. As noted, there are some protists for which developing a durable cyst form is one of the possibilities. To address this, researchers often need to sequence various life stages and use morphological, ecological, and molecular information to assess species accurately. The existence of cryptic species, which are visually indistinguishable but genetically distinct, also poses a challenge to DNA barcoding, as shown in Table 4.1. This problem can be addressed by using highly accurate molecular markers and extensive sampling (Hoef-Emden, 2012).

Species Identification: DNA barcoding helps accurately identify protistan species, especially morphologically similar ones. It also differentiates species within genera such as *Plasmodium*, *Giardia*, and *Trypanosoma*.

Biodiversity Assessment: Facilitates rapid assessment of protistan diversity in various environments, including marine, freshwater, and soil. Estimating the diversity of diatoms, ciliates, and flagellates in environmental samples.

Environmental Monitoring: DNA barcoding can detect protistan species in environmental samples, which is useful in monitoring ecosystem health. It can also be used to monitor water quality by identifying harmful algal blooms caused by species like *Karenia brevis*.

Phylogenetic Studies: This field helps construct phylogenetic trees and understand the evolutionary relationships among protists. It also resolves the phylogeny of groups such as foraminifera, radiolarians, and dinoflagellates.

Biogeography: This field aids in studying the distribution patterns of protistan species across different geographical regions. It is also used to investigate the global distribution of protists like *Emiliania huxleyi* in marine ecosystems.

Discovery of New Species: Enables the discovery and documentation of new protistan species that are difficult to distinguish morphologically. Identification of new species in genera like *Cryptomonas* and *Chlorella* through DNA barcoding efforts.

Pathogen Identification: DNA barcoding is used to identify protistan pathogens responsible for human, animal, and plant disease. Identifying species of *Leishmania*, *Trypanosoma*, and *Plasmodium* as causative disease agents.

Food Web Studies: This helps trace the role of protists within food webs by identifying species in the gut contents of predators. It also analyzes the diets of marine zooplankton by identifying ingested protists.

Ecological and Functional Studies: Assists in linking protistan diversity to ecological functions and their role in nutrient cycling. Understanding the role of ciliates in nitrogen cycling in aquatic ecosystems through species identification.

Conservation Biology: Used in conservation efforts to protect biodiversity by identifying key protistan species in various habitats. It also identifies endangered protistan species in coral reef ecosystems and freshwater environments.

Technically, DNA extraction, polymerase chain reaction, and sequencing present challenges to the effectiveness of DNA barcoding in protists. One of the reasons for this is that the cells of protists can be challenging to break open, and the DNA contained in them might be vulnerable to degradation or contamination by other organisms in environmental samples. Thus, ensuring that the DNA extraction protocol is optimal and minimizes contamination with substances other than protist cells is important. In addition, a significant challenge of DNA barcoding in protists using universal markers has been designing primers that would bind only to the DNA of protists, not to the DNA of their mitochondria or chloroplasts, while maintaining isolation from foreign sequences. It is also possible for the barcode markers to contain introns, known for their poor PCR amplification and unpredictable sequence alignment, or other regions of high variability in their sequences. In either case, the challenge will have to be managed by limited primer design, and reliable bioinformatics tools will need to be used for data analysis (Gamit and Amin, 2024).

8.6 DNA BARCODING IN PROTISTS

Many studies have shown the efficiency of DNA barcoding in protist studies and other studies. In particular, it allows us to identify new species, explore protist diversity in various ecosystems, and monitor environmental changes. The following case studies illustrate the broad range of uses and importance of DNA barcoding in protist studies.

8.6.1 Identification of New Species

One of the most interesting uses of DNA barcoding is discovering unknown and previously undescribed protist species. Traditional approaches to this issue rely upon morphological approaches that tend to miss out on hidden species and subtle differences. At the same time, DNA barcoding allows the identification of previously unknown species based on the differences observed at the genetic level. For example, a case study related to marine dinoflagellates was conducted using the DNA barcode marker COI and NTS to highlight multiple cryptic and previously undescribed species of these organisms. The discovered hidden species largely differ from known species in their genetic structure. These findings have important applications for understanding marine biodiversity and the ecosystems of different dinoflagellate species. In freshwater ecosystems, there have also been discovered hidden species of diatoms based on the SSU rRNA gene and the *rbcL* gene. Often, the life of such organisms depends on the proper functioning of primary producers, such as diatom species. New species discovery helps to understand ecosystems in rivers and lakes better. Establishing new protist species through DNA barcoding also has important implications for biogeography and land conservation (Letsiou et al., 2024).

8.6.2 Exploring the Richness of Protist Diversity in Various Ecosystems

DNA barcoding has greatly enhanced our understanding of protist diversity across various ecosystems. Protists are crucial in primary production, nutrient cycling, and food web dynamics of marine environments. DNA barcoding studies have elucidated the diversity of aquatic protists, namely phyto—and zooplankton, and how these species are distributed across various sampling locations and periods. For instance, barcoding in metagenomic studies of seawater samples revealed many protist communities in different oceanic areas. The survey of del Campo et al. pointed to the drivers of changes in protists' quantitative and qualitative composition. DNA barcoding studies in terrestrial ecosystems utilized soil protists, which have similar roles to those in marine communities. Various soil sample analyses, such as amoeba, flagellates, and ciliates, have demonstrated species diversity. The studies also revealed a complex character of soil protist communities and their relation to other microorganism groups, including bacteria and fungi. Gaining knowledge of the diversity of protists in the soil is crucial to maintaining sustainable land management and agriculture, since the contribution of these microorganisms is significant for soil fertility and plant health promotion (Caron et al., 2012).

In freshwater ecosystems, DNA barcoding has provided helpful information regarding the diversity and dynamics of protist communities in rivers, lakes, and wetlands. Freshwater protists are critical to microbial food webs, nutrient cycling, and energy flow. DNA barcoding studies have detected a range of freshwater protists, including those free-living protists and those in symbiosis. These data demonstrate the importance of freshwater protists to the functioning of ecosystems, hence the necessity to preserve and protect freshwater biodiversity (Hadi et al., 2016).

8.6.3 Monitoring Environmental Changes and Impacts

DNA barcoding provides a powerful tool for monitoring environmental changes and assessing their impact on protist communities. Protists are sensitive to alterations in their environment and, as such, can serve as good indicators of the overall health of an ecosystem. By analyzing protist diversity and composition, it is possible to detect early signs of stress and degradation of the environment (Hajibabaei et al., 2011).

A good example is monitoring harmful algal blooms in marine and freshwater systems. HABs are usually caused by, among other factors, dinoflagellates and cyanobacteria. These blooms can lead to severe ecological and economic implications. DNA barcoding helps identify species that form the bloom automatically and promptly, which aids in mitigating and managing the problem. For example, during blooms in bodies of water, toxin-producing species have been identified by metagenomic analyses of water samples. The information thus obtained helped devise an efficient plan for monitoring and managing the problem (Liu et al., 2022).

Apart from that, DNA barcoding has been applied to assess pollution and habitat destruction impacts on protist communities. In polluted environments, such as industrial wastewater and agricultural runoff, DNA barcoding is used to determine changes in protist diversity and identify the species that can tolerate pollution and the ones sensitive to it. These data should be used to assess the efficacy of pollution management measures and how to restore and preserve damaged ecosystems (Jacobs-Palmer, 2021).

Climate change is one of the areas where DNA barcoding has proved its effectiveness. Both fluctuating temperatures, ocean acidification, and altered precipitation potentially impact the diversity of protist communities and their important functions in different ecosystems. Some studies on DNA barcoding have traced the changes in protist diversity and distribution due to the factors related to climate change. For example, in both the Arctic and Antarctica, the retreat of ice cover and the warming of waters detected through DNA barcoding have affected the changes in protist communities. These findings have important implications for understanding engineering prospects and mitigating climate change's effects on polar ecosystems (Walczynska et al., 2018).

DNA barcoding is now commonly used as a reliable and unbiased tool for studying protist diversity and its associated challenges. The tool allows for an accurate and time-efficient way of identifying species in a particular ecosystem and conducting the required research to count them. Even though they are more genetically diverse than other eukaryotic cells and have a wide variety in their complex life cycles, DNA barcoding has enabled scientists to track and catalog the diversity of protists effectively. On the one hand, this has further supported the complex nature of the group (Pawlowski et al., 2012). On the other hand, unable to see the whole ranges of these species and benefiting from a significant support mechanism invisible through a microscope, humanity has taken steps to protect protists and their functions in different environments. It is evident in real-life situations and their practical activities, which are discussed in the paper. In the context of the growing facilitation of the stimulus provided by the untapped molecular tools presented to the crew, the importance of DNA barcoding in realizing the underestimated diversity of protists and managing and conserving them to promote or maintain the vital functions of these small organisms and their populations will only grow (Gostel and Kress, 2022).

8.7 CHALLENGES IN EVALUATING PROTISTS' BIODIVERSITY

There are several difficulties in evaluating protist biodiversity due to their extensive genetic diversity, ability to change shape, hidden species, and intricate life cycles. Understanding these challenges can help develop suitable approaches to analyze and protect protist biodiversity, which is vital for the existence and well-being of ecosystems. For instance, vast genetic diversity causes problems in categorizing and understanding protist richness. Genetic diversity refers to the species' genetic code's full range of genetic characteristics. Modern protists are incredibly diverse, seen at all levels: gene aggregates, gene profiles, and other molecular aspects. In some cases, there are quite a few distinct genotypes in a population that, in combination, make up a single species. This genetic diversity can lead to significant morphological, physiological, and ecological differences and, as a result, cause difficulties in recognizing the boundaries of new and existing species and their evolutionary relationships (Barbosa, 2022).

The companion characteristic of extensive genetic diversity is the ability of a single species to exhibit a high diversity of physical forms in response to different environmental conditions. Thus, the plasticity of unicellular protists may be so high that individuals from the same species may have significant differences. As a result, various forms of a single protist may be misidentified, and, in some cases, several forms may be classified as distinct species. For example, protists can change their shape and size under certain conditions. As a result of this transformation, they can vary depending on factors such as the presence of predators, temperature, availability of nutrients, and pH in the environment. Considering these features, these organisms inhabit a wide range of environmental conditions, which makes it impossible to classify them according to their physical structure reliably. One of the key ways to address these challenges is to use molecular approaches, such as DNA barcoding, to provide a genetic basis for recognizing new and existing species. However, this technique may face problems with extensive genetic heterogeneity that can be identified between the same populations of protists. Indeed, taxonomy depends on comparing DNA sequences of different species or isolates and measuring genetic distances. Many alleles and genetic variants can compromise this process.

In addition, horizontal gene transfer is widespread among modern protists and can complicate genetic distinction. Moreover, morphological plasticity is responsible for additional difficulties that must be considered while studying biodiversity and genetic extent (Mulot et al., 2017). At times, protists can change their shape in response to environmental conditions. For example, some amoeboid protists can change their shape and the way they move when food sources are available or when there are predators. On top of that, based on physical characteristics, it is often difficult to track changes in protists and provide taxonomists with an effective classification system (Barbosa, 2022).

Cryptic species present a major problem when it comes to the assessment of protist biodiversity. A cryptic species is a group of organisms that might be genetically unique but appear indistinguishable to an observer. Both the successions of environmental changes occur in the phylogenetically and its proxy. Body sequences derived from specific groups may also share uniform appearances and might not be adequately differentiated within modern taxonomic techniques, which are mainly based on morphology; examples include many groups of protists, in which cryptic species are common, making them incredibly difficult to distinguish, to measure biodiversity accurately, and to obtain information on the ecological interactions between them. When evaluating such organisms, molecular methods are adopted to identify cryptic species, such as DNA barcoding, metagenomics, and phylogenetic analyses, which are used to detect genetic variations pharmacological observations alone might not discover (Chenuil, 2019).

However, using molecular methods when identifying cryptic species is also troublesome. If the identification of cryptic species is needed, it depends primarily on selecting the best genetic markers, for several markers might not be precise enough to differentiate closely related species. Additionally, when determining the criteria for analyzing genetic data and regulating their simplicity, definitions of the species limits may vary case by case, depending on the species group and the study context [58]. Finally, protists have many different life stages that can be distinguished, for these life stages reflect a unique form and function and appear very different. In addition, the protists' different life stages reproduce in various ways and occupy different ecological niches, resulting in multiple forms. They reproduce in a variety of ways. Some protists create cysts, a dormant phase that allows them to survive extreme environmental conditions. In the cyst stage, protists may appear and function differently than active vegetative stages. Some protists follow complex reproductive cycles that include both sexual and asexual phases, which are separated in form and function. To understand such life cycles properly, full and careful observation and correct sampling of different conditions and periods must be taken. Moreover, protist populations fluctuate in great numbers due to varying conditions of time and space at different periods of the year, including changes in seasons, environmental disruptions, and other organisms. As a result, long-term studies have eventually been necessary to change and assess the patterns of the populations. It can be a challenging process based on the protozoa found in some of the regions and conditions where they are common (Corliss, 2002).

The ecological roles of protists create difficulties in the assessment of biodiversity complexity. First, they are ubiquitous and variable, as they have been identified in various ecological niches. Second, they are involved in multiple environmental functions, such as primary producers, decomposers, symbionts, and parasites. The relationships of protists with other bacteria, plants, animals, and fungi can drastically vary and be complex. These relationships can also impact the distribution and abundance of specific protist species and their genetic diversity and morphological characteristics. These issues imply that studying the ecological roles of protists requires a spectrum of disciplines, such as ecology, genetics, and microbiology. The impact of environmental change and human action can be significant but complex. Pollution, climate change, habitat destruction, and introduction of invasive species are some environmental changes and human actions affecting protist diversity. It implies that the response to the current challenges should be based on an understanding of these processes and their consequences and should be applied in conservation (Carvalho da Silva and Fernandes, 2023).

However, molecular approaches can provide an adequate response to assessing genetic diversity, which is often not fully evaluated by morphological methods, those of identifying cryptic species, and those that have a complex life cycle. Thus, combining conventional and contemporary approaches is required to address the issues and pitfalls mentioned above. Overall, assessing protist biodiversity is a complex and challenging task, as these microorganisms are very heterogeneous in terms of genetic variations, physical characteristics, presence of hidden and cryptic species, and life cycle complexities. However, these issues can be addressed by combining conventional and contemporary approaches, undertaking comprehensive sampling, and investigating the latter to address the challenges. The study of protist biodiversity can help to gain a deeper understanding of these microorganisms, their ecological roles, and the factors contributing to their distribution and abundance (Hawksworth, 2009).

8.8 TRENDS IN DNA BARCODING

The interactive process underlying the development of DNA barcoding technology has remarkably moved on and offers better means and methods for covering biodiversity. Advanced techniques enhance DNA barcoding's precision, efficiency, and universe, especially its use for protists. Also, other molecular and bioinformatics tools may operate together with DNA barcodes, turning them into new data and deployment sources. Such advancements possess the potential to extend our understanding of protist diversity and have even broader possibilities for different environmental, conservational, and large-scale biodiversity assessment purposes.

This chapter describes how recent trends in DNA barcoding technology development greatly benefit its efficiency and expand the field of its use. It is important to state that one of the crucial advancements is the enhancement related to sequencing technologies. NGS, or next-generation sequencing platforms, are the most significant recent contributors to the development of DNA barcoding. It is especially relevant to Illumina, Ion Torrent, and Pacific Biosciences. They all have advantages regarding high throughput, better accuracy, and costs. Indeed, such platforms may produce massive amounts of sequence data for a comparably short period, which has never been possible. As a result, studying biodiversity in a much larger universe is beneficial, involving multiple samples and barcode regions in the process, subsequently enhancing the depth, resolution, and scope of particular applications (Grant et al., 2021).

While the COI gene is the most frequently used barcode for animals, followed by its mitochondrial attribution process, many other markers exist, and some of them are optimized for use with numerous groups of organisms, including protists. Examples are the Internal Transcribed Spacer, or ITS, region, Small Subunit Ribosomal RNA, or SSU rRNA, and transcribed as rbcL, the large subunit of rbcL, or RuBisCo. They all allow for species separation with outstanding precision, and many are applicable over wide taxonomic ranges (Pentinsaari et al., 2016).

There have been numerous advances in DNA barcoding, and bioinformatics tools and databases have facilitated several of them. The Barcode of Life Data System and NCBI GenBank are two examples of a centralized repository of barcode sequences and associated information. They have significantly improved the researchers' ability to share data with others and perform comparative analyses. In addition, state-of-the-art bioinformatics tools, including QIIME, USEARCH, and MEGA, allow for efficient data processing by comparing and sorting large datasets of sequences into groups, performing sequence alignment and phylogeny construction, and determining genetic distances between species. Integrating bioinformatics tools analyzes sequence data more efficiently, making species identification or phylogenetic analysis more precise, quick, and cost-effective (Mathew, 2013).

Integrative DNA Barcoding: Due to the integration of DNA barcoding with other molecular and bioinformatics tools, the spectrum of its application has been expanded, and the method's reliability has been significantly increased. For example, integrating DNA barcoding and metagenomics is an interesting and relevant approach to comprehensively studying microbial populations. Metagenomics is a technique that involves sequencing the collective genome of all organisms in a single environmental sample. DNA barcoding and metagenomics integration allow for comprehensive analysis of microbial communities, including protists, bacteria, and archaea, without the need for culture. This technique makes it possible to identify the presence of rare or hard-to-culture species, discover hidden diversity, and provide insight into the community's composition and functional capacity. Transcriptomics, the study of RNA transcripts, is another molecular tool that can be combined with DNA barcoding. Using RNA sequencing, it is possible to determine the specific combinations of genes that are expressed or not in the cells of the organisms and, thus, predict the functions and adaptations of these species to their environment (Chesters et al., 2015).

Environmental DNA analysis: eDNA analysis is an exciting field using DNA barcoding to detect and identify different organisms in various environmental samples, including water, soil, and air. It is highly valuable for biodiversity monitoring in aquatic ecosystems, where other methods can be challenging. The eDNA sequencing allows us to track a wide range of species in a given area, even if they are rare or hard to find. As a promising tool, eDNA analysis has an important implication for conservation, investigated by Jurado-Rivera, Pons, Alvarez, and others. It allows for monitoring endangered species, discovering the presence of invasive species, and assessing ecosystem health (Nørgaard et al., 2021).

8.9 EMERGING TRENDS AND POTENTIAL IN PROTIST RESEARCH

Focusing on ecology might not be the only possible application of DNA barcoding. It could also be used in public health, agriculture, and industry. For example, in public health, DNA barcoding could help identify pathogenic protists and track the spread of some infectious diseases. In agriculture, it could be used to determine protist species, particularly parasites, that might impact plants, thus contributing to food security. In industry, it could help maintain quail control; for instance, DNA barcoding could enable companies to check that bio-based products are not contaminated by other organisms and assess that the microbial strains used for fermentation are appropriate. Overall, some of the most interesting possibilities for the future are closely tied to the interaction between scientists and the public. On the one hand, using DNA barcoding for corresponding citizen science initiatives would enable researchers to expand their sampling to new territories. On the other hand, the approach effectively raises the public's awareness about the issue.

Advancements in portable and easy-to-use sequencing technologies, such as the Oxford Nanopore MinION, are being made continuously and show significant promise for DNA barcoding in the field. At this point, real-time DNA sequencing and analysis with the help of portable sequencing devices in remote field locations allows researchers to conduct biodiversity assessments and identify species in situ. Portable sequencing technologies are also beneficial in addressing environmental issues, such as detecting harmful algal blooms and investigating the impacts of ecological disasters. Overall, technological advancements in DNA barcoding and its integration with other molecular tools and bioinformatics greatly increase our power to study and understand the diversity of protists. The potential future applications of DNA barcoding continue to pose numerous opportunities to investigate protist diversity and its changes over time to preserve and conserve protist species and their environments. As sequencing technologies continue to advance and become more accessible, the importance of DNA barcoding in revealing the hidden diversity of protists and guiding conservation in their ecosystems will continue to grow. Thus, the cooperation between researchers, conservationists, policymakers, and the general public will play an essential role in realizing the full potential of DNA barcoding in biodiversity research and conservation.

8.10 SOCIOLOGICAL DIMENSIONS OF DNA BARCODING IN PROTIST RESEARCH

DNA barcoding of protists is a phenomenon of not just biological or technological improvement but also extreme sociological significance. It has implications in the way societies generate knowledge; value biodiversity; govern the environment; and negotiate power, equity, and the well-being of the citizens. One of the sociological aspects relates to politics of knowledge production. DNA barcoding is also transforming the face of scientific discovery, even as laboratories in the Global South start supplying the world with biodiversity depositories. This democratization of microbial studies attempts to disrupt Western hegemony in biodiversity studies and enables local scientists, students, and institutions to establish their scientific sense of agency/identity. The other aspect of interest is the way in which the societies recognize and attach importance to the invisible biodiversity. Protists are microscopic and thus they are not part of the cultural imagination, the education and conservation discourse. Knowledge of these perceptions is of importance since DNA barcoding can change the consciousness of the people since it has the capacity to transform microbial diversity to make it visible, culturally significant and of significance to environmental management.

Sociological insight is also helpful in environmental governance and conservation planning. The data of DNA barcoding may be useful in policy making concerning soil health, monitoring freshwater ecosystems, land-use planning, and conservation priorities, yet the institutionalization of scientific evidence is profoundly influenced by political interests, bureaucratic systems, and the social aspects of decision-making. Sociologists assist in the uncovering of how the scientific suggestions are implemented—or even not implemented at all—within the governance systems. Further, the protist-dominated habitats like lakes, wetlands, and soils have a strong connection with the daily life and ecological experience of the local people. The sociological perspective brings to the fore the alignment of indigenous or community-based ecological knowledge with molecular data and therefore provides the opportunity to develop co-created conservation policies that take into consideration the local viewpoints, as well as providing equitable access to natural resources. By so doing, DNA barcoding can be used to facilitate more comprehensive environmental management.

Socio-economic significance of protist biodiversity studies is also notable. The protists are vital to the agricultural productivity, soil fertility, nutrient cycling, fisheries, and human health. The DNA barcoding can allow identification and tracking of dangerous or useful species within a short time, which can impact farming, water management, and disease control. Sociologists study the meaning, acceptance, or rejection of these technological interventions by various social groups, such as farmers, fishers, local authorities, and policymakers, depending on their economic interests, cultural values, and their scientific literacy. This is connected to ethical and legal considerations. DNA barcoding has been associated with issues on the ownership of genetic resources concerning genetic resource sovereignty, data sovereignty and bioprospecting in a variety of frameworks, including the Nagoya Protocol. The sociological analysis assists in studying the question of who owns the microbial genetic resources, who gains the profit of commercialization of microbial diversity, how trust is established between the scientists and the community and how it is possible to avoid exploitative research methods.

Another important sociologically significant dimension is public health, in particular, due to some protists being associated with disease. Although barcoding enhances the early identification of parasitic or waterborne pathogens, the response of the society relying on the risk communication and cultural beliefs of the disease and the community health behaviors. The inclusion of sociology will find application whereby the monitoring of microbes translates to effective education of the people on health and at the community level. Lastly, the DNA barcoding reveals scientific disparities in the world. Most low- and middle-income nations do not have sequencing facilities, training, and funding. Sociologists emphasize such inequalities and advocate the principles of building their capacity, fair partnerships, and policies that can avoid bidirectional dismissal of biodiversity data to high-income countries. In that regard, DNA barcoding of protists has not only an ecological but also a fundamentally social future to it; it overlaps with development policies, environmental justice, and the agenda of more inclusive scientific futures.

REFERENCES

Adams, J., Transcriptome: connecting the genome to gene function. *Nature Education*, 2008. **1**(1): p. 195.

Adl, S.M., et al., The revised classification of eukaryotes. *Journal of Eukaryotic Microbiology*, 2012. **59**(5): p. 429–514.

Ahmed, N.H., Cultivation of parasites. *Tropical Parasitology*, 2014. **4**(2): p. 80–89.

Ares, M.A., et al., The coli surface antigen CS3 of enterotoxigenic escherichia coli is differentially regulated by H-NS, CRP, and CpxRA global regulators. *Frontiers in Microbiology*, 2019. **10**.

Barbosa, M., et al., Trophic state drives the diversity of protists in a tropical river (New River, Belize). *Microorganisms*, 2022. **10**(12).

Barcaccia, G., M. Lucchin, and M. Cassandro, DNA barcoding as a molecular tool to track down mislabeling and food piracy. *Diversity*, 2016. **8**(1): p. 2.

Burki, F., M.M. Sandin, and M. Jamy, Diversity and ecology of protists revealed by metabarcoding. *Current Biology*, 2021. **31**(19): p. R1267–R1280.

Caron, D.A., et al., Marine protistan diversity. *Annual Review of Marine Science*, 2012. **4**: p. 467–493.

Carvalho da Silva, V. and N. Fernandes, Protist taxonomic and functional diversity in aquatic ecosystems of the Brazilian Atlantic forest. *PeerJ*, 2023. **11**: p. e15762.

Chandarana, K.A. and N. Amaresan, Soil protists: an untapped microbial resource of agriculture and environmental importance. *Pedosphere*, 2022. **32**(1): p. 184–197.

Cheng, Z., et al., The devil is in the details: problems in DNA barcoding practices indicated by systematic evaluation of insect barcodes. *Frontiers in Ecology and Evolution*, 2023. **11**: p. 1149839.

Chenuil, A., et al., Problems and questions posed by cryptic species: a framework to guide future studies. In *From Assessing to Conserving Biodiversity: Conceptual and Practical Challenges*, E. Casetta, J. Marques da Silva, and D. Vecchi, Editors. Springer International Publishing, Cham, 2019. p. 77–106.

Chesters, D., et al., A DNA barcoding system integrating multigene sequence data. *Methods in Ecology and Evolution*, 2015. **6**(8): p. 930–937.

Chheda, U., et al., Factors affecting stability of RNA–temperature, length, concentration, pH, and buffering species. *Journal of Pharmaceutical Sciences*, 2024. **113**(2): p. 377–385.

Corliss, J.O., Biodiversity and biocomplexity of the protists and an overview of their significant roles in maintenance of our biosphere. *Acta Protozoologica*, 2002. **41**(3): p. 199–220.

Cowan, R.S. and M.F. Fay, Challenges in the DNA barcoding of plant material. *Methods in Molecular Biology*, 2012. **862**: p. 23–33.

Dawan, J. and J. Ahn, Application of DNA barcoding for ensuring food safety and quality. *Food Science and Biotechnology*, 2022. **31**(11): p. 1355–1364.

Dawson, S. and L. Fritz-Laylin, Sequencing free-living protists: the case for metagenomics. *Environmental Microbiology*, 2009. **11**: p. 1627–1631.

Deshpande, D., et al., RNA-seq data science: from raw data to effective interpretation. *Frontiers in Genetics*, 2023. **14**.

El-Diasty, M., et al., Isolation of Brucella abortus and Brucella melitensis from seronegative cows is a serious impediment in Brucellosis control. *Veterinary Sciences*, 2018. **5**(1).

Elsaied, H., et al., Applications and challenges of DNA barcoding and metabarcoding in African fisheries. *Egyptian Journal of Aquatic Research*, 2021. **47**(1): p. 1–12.

Firkins, J.L., et al., Extending Burk Dehority's perspectives on the role of ciliate protozoa in the rumen. *Frontiers in Microbiology*, 2020. **11**: p. 123.

Fournier, B., et al., Toward the use of protists as bioindicators of multiple stresses in agricultural soils: A case study in vineyard ecosystems. *Ecological Indicators*, 2022. **139**: p. 108955.

Gamit, A. and D. Amin, DNA barcoding techniques for protists. In *Practical Handbook on Soil Protists*. Springer, 2024. p. 165–173.

Gostel, M.R. and W.J. Kress, The expanding role of DNA barcodes: indispensable tools for ecology, evolution, and conservation. *Diversity*, 2022. **14**(3): p. 213.

Grant, D.M., et al., The future of DNA barcoding: reflections from early career researchers. *Diversity*, 2021. **13**(7): p. 313.

Hadi, S.I., et al., DNA barcoding green microalgae isolated from neotropical inland waters. *PLoS One*, 2016. **11**(2): p. e0149284.

Hajibabaei, M., et al., Environmental barcoding: a next-generation sequencing approach for biomonitoring applications using river benthos. *PLoS One*, 2011. **6**(4): p. e17497.

Hawksworth, D.L., *Protist Diversity and Geographical Distribution*. Springer Netherlands, 2009.

Hebert, P.D.N. and T.R. Gregory, The promise of DNA barcoding for taxonomy. *Systematic Biology*, 2005. **54**(5): p. 852–859.

Hebert, P.D.N., et al., Biological identification through DNA barcodes. *Proceedings of the Royal Society of London B*, 2003. **270**: p. 313–321.

Hoef-Emden, K., Pitfalls of establishing DNA barcoding systems in protists: the cryptophyceae as a test case. *PLoS One*, 2012. **7**(8): p. e43652.

Jacobs-Palmer, E., et al., Environmental DNA metabarcoding for simultaneous monitoring and ecological assessment of many harmful algae. *Frontiers in Ecology and Evolution*, 2021. **9**.

Jalali, S. and R. Ojha, *DNA Barcoding for Identification of Agriculturally Important Insects*. Springer, 2015. p. 13–23.

Krehenwinkel, H., A. Pomerantz, and S. Prost, Genetic biomonitoring and biodiversity assessment using portable sequencing technologies: current uses and future directions. *Genes (Basel)*, 2019. **10**(11).

Kress, W.J. and D.L. Erickson, DNA barcodes: methods and protocols. *Methods in Molecular Biology*, 2012. **858**: p. 3–8.

Lahaye, R., et al., DNA barcoding the floras of biodiversity hotspots. *Proceedings of the National Academy of Sciences*, 2008. **105**(8): p. 2923–2928.

Lebret, K., et al., Genetic diversity within and genetic differentiation between blooms of a microalgal species. *Environmental Microbiology*, 2012. **14**(9): p. 2395–2404.

Letsiou, S., et al., DNA barcoding as a plant identification method. *Applied Sciences*, 2024. **14**(4): p. 1415.

Liu, C., X. Zhang, and X. Wang, DNA metabarcoding data reveals harmful algal-bloom species undescribed previously at the Northern Antarctic Peninsula region. *Polar Biology*, 2022. **45**(9): p. 1495–1512.

Mathew, D., *DNA Barcoding and Its Applications in Horticultural Crops*. New India Publishing Agency, New Delhi, 2013. p. 25–50.

Mathur, R., et al., DNA barcoding for identification and detection of species. *Letters in Applied NanoBioScience*, 2021. **11**: p. 3542–3548.

Mulot, M., et al., Genetic determinism vs. phenotypic plasticity in protist morphology. *The Journal of Eukaryotic Microbiology*, 2017. **64**(6): p. 729–739.

Mulvaney, J., M. Moir, and M.I. Cherry, DNA barcoding reveals cryptic diversification and taxonomic discordance among bats and birds within Sub-Saharan Africa. *Biodiversity and Conservation*, 2023. **32**(14): p. 4895–4914.

Muñoz, J.F., et al., Genome analysis reveals evolutionary mechanisms of adaptation in systemic dimorphic fungi. *Scientific Reports*, 2018. **8**(1): p. 4473.

Nørgaard, L., et al., eDNA metabarcoding for biodiversity assessment, generalist predators as sampling assistants. *Scientific Reports*, 2021. **11**(1): p. 6820.

Nurhasanah, S. and N. Papuangan, Amplification and analysis of Rbcl gene (Ribulose-1, 5-Bisphosphate Carboxylase) of clove in Ternate Island. In *IOP Conference Series: Earth and Environmental Science*. IOP Publishing, 2019.

Odah, M.A.A., Unlocking the genetic code: exploring the potential of DNA barcoding for biodiversity assessment. *AIMS Molecular Science*, 2023. **10**(4): p. 263–294.

Pawlowski, J., et al., CBOL protist working group: barcoding eukaryotic richness beyond the animal, plant, and fungal kingdoms. *PLoS Biology*, 2012. **10**(11): p. e1001419.

Pentinsaari, M., et al., Molecular evolution of a widely-adopted taxonomic marker (COI) across the animal tree of life. *Scientific Reports*, 2016. **6**(1): p. 35275.

Ratnasingham, S. and P.D.N. Hebert, A DNA-based registry for all animal species: the barcode index number (BIN) system. *PLoS One*, 2013. **8**(7): p. e66213.

Silva, F.C.D., et al., Yellow fever vaccination in a mouse model is associated with uninterrupted pregnancies and viable neonates except when administered at implantation period. *Frontiers in Microbiology*, 2020. **11**.

Smith, M.A., B.L. Fisher, and P.D. Hebert, DNA barcoding for effective biodiversity assessment of a hyperdiverse arthropod group: the ants of Madagascar. *Philosophical Transactions of the Royal Society B: Biological Sciences*, 2005. **360**(1462): p. 1825–1834.

Stein, E.D., et al., Evaluating ethanol-based sample preservation to facilitate use of DNA barcoding in routine freshwater biomonitoring programs using benthic macroinvertebrates. *PLoS One*, 2013. **8**(1): p. e51273.

Tate, D.F., et al., Sustaining eHealth engagement in a multi-year weight gain prevention intervention. *Obesity Science & Practice*, 2019. **5**(2): p. 103–110.

Triplett, L.R., S.J. Taerum, and R.R. Patel, Protists at the plant-bacterial interface: impacts and prospective applications. *Physiological and Molecular Plant Pathology*, 2023. **125**: p. 102011.

Walczynska, K., M. Mańko, and A. Weydmann-Zwolicka, *Arctic Ocean Biodiversity and DNA Barcoding—a Climate Change Perspective*. Springer, 2018. p. 145–153.

Weydmann-Zwolicka, A., et al., Comparison of DNA metabarcoding and microscopy in analysing planktonic protists from the European Arctic. *Marine Biodiversity*, 2024. **54**(3): p. 44.

Whittaker, R.H. and L. Margulis, Protist classification and the kingdoms of organisms. *Biosystems*, 1978. **10**(1–2): p. 3–18.

Yu, N., Y. Chen, and Y. Li, Chapter 12—aptamer-based characterization of food authenticity and nutrition. In *Aptamers for Food Applications*, Y. Dong, Editor. Academic Press, 2023. p. 277–292.

9 DNA Barcoding of Plant Parasitic Nematodes

Status and Prospects

Muhammad Amjad Ali, Malaika Tariq, Usman Arshad, Abdelfattah A. Dababat, Cemal Kurt, Ikhlas Shafique, Muhammad Jabran, and Amjad Abbas

9.1 INTRODUCTION

Plant-parasitic nematodes (PPNs) belong to the group of non-segmented, small roundworms that infect plant roots and cause significant losses in agricultural productivity. They cause $152 billion in losses worldwide every year, infecting a wide range of crops and contributing to concerns about food security (Dababat et al., 2025). The complication in managing PPNs is that they can inhabit a broad variety of environments and exhibit several feeding habits, including both migratory and sedentary behaviors. PPNs are highly diverse, comprising over 4,100 species with various host preferences and life cycles. Identification of PPNs is crucial for implementing effective control measures, as different species require specialized management strategies (Kantor et al., 2024). Nematodes are regarded as the most challenging animals to differentiate. Due to their high intra-specific morphological variation, these species are very difficult to identify using morphometric studies and conventional morphological approaches, which can result in ambiguous evaluations and extensive overlap among many characters. Additionally, PPNs' symptoms are similar to those caused by other biotic stresses, such as drought and frost, making them even more challenging to identify. Therefore, new methods are needed to ensure accurate species identification (Bogale et al., 2020; Jabbar et al., 2020).

Nematode identification relying on conventional methods is time-consuming for morphological and morphometric analysis of multiple samples (Arshad et al., 2022). The reliability and accuracy of this type of identification depend mainly on the expertise of skilled and experienced personnel in making the diagnosis. However, the number of experienced and qualified nematode taxonomists is small and currently declining. Therefore, rapid and accurate methods are necessary to identify plant pathogens and pests, ensuring safe, healthy, and increased food production. To this end, technology is constantly being updated, and new methods are emerging to identify and manage the problem. Molecular methods for the identification of nematodes provide accurate substitute diagnostic approaches. In 1990, the discovery of

DOI: 10.1201/9781003538929-9

PCR-based species identification techniques led to the gradual development of the concept of a standardized molecular identification system. Molecular identification has been widely employed for the routine diagnosis of pathogenic strains and surveys of microbial diversity, as culture-independent identification technology is required (Shao et al., 2023).

The recent, rapid, and efficient technique for PPN identification is DNA barcoding. DNA barcoding, a molecular approach that employs short, standardized DNA sequences as identifiers of species, has transformed the discipline of nematode taxonomy and diagnostics. By providing a swift, precise, and low-cost approach to species identification, DNA barcoding has emerged as an invaluable tool for researchers and practitioners alike (Antil et al., 2023). DNA barcodes, which consist of a brief, uniform DNA sequence of 400–800 bp, should be easily identifiable and characterized for all land-based species (Savolainen et al., 2005). Referring to the mouse mitochondrial genome, the DNA barcode itself consists of a 648 bp region, 58–705, from the 5′ end of the COI (cytochrome c oxidase I) gene. The premise of this research is that every species will likely possess a unique DNA barcode. According to estimates, there are approximately 4,650 possible ATGC combinations out of an estimated 10 million yet-to-be-discovered species. Furthermore, it is believed that between-species genetic diversity is greater than within-species (Frézal & Leblois, 2008). DNA barcoding particularly aims to (i) identify unknown individuals to a particular species and (ii) improve the identification of novel species, especially in cryptic, microscopic, and other organisms with complicated or inaccessible morphology (Odah, 2023).

9.2 CONVENTIONAL AND MODERN METHODS OF NEMATODE IDENTIFICATION

Since the 19th century, nematodes have been classified according to their morphological and morphometric characteristics. The physiological characteristics of the reproductive organs of mature male nematodes are sometimes used to identify species. Although the main advantage of this method is that it is almost free of cost, its disadvantages include the need for considerable knowledge and training, inability to identify specimens at the species level if the taxonomist only has access to eggs or larvae, or if multiple specimens are required to collect sufficient morphometric data to allow a confident classification of the nematode species under study.

For example, the root-knot nematode (Meloidogyne spp.) has been identified based on a female perineal pattern, specifically the tail terminus, which surrounds the cuticular striae, lateral lines, phasmids, and the perineum (vulva-anus area) in the posterior region. These characteristics were previously used to distinguish and identify different species of Meloidogyne (*M. incognita*, *M. hepla*, *M. javanica*, and *M. arenaria*) (Karssen & Van Aelst, 2001; Anwar et al., 2022). However, upon the identification of new species, these perineal patterns became insufficient for the identification of nematodes, as the perineal patterns and other morphological characters overlapped between species (Villar-Luna et al., 2016). Decades later, other research studies highlighted that morphological characterization was insufficient to classify nematodes. Different methods for identifying root-knot nematodes have been implemented in recent years, utilizing the potential of combining both molecular and morphological characters (Ye et al., 2019; Jabbar et al., 2020). Other scientists proposed reordering nematodes within a homogeneous phylogroup based on 18S ribosomal RNA sequences, as determined by PCR (Mullis et al., 1986).

In addition, several varieties of polymerase chain reaction (PCR) methods are also used in molecular characterization. While the RFLP (Restriction Fragment Length Polymorphism) method has the potential for precision, sensitivity, and speed (Jabran et al., 2023), this method has been successfully utilized in many cases to identify root-knot nematodes and their physiological subspecies, such as *Bursaphelenchus* spp. (Aikawa et al., 2013) and Xiphinema americanum. Currently, this approach is not widely applied due to technical difficulties and the requirement of large amounts of targeted DNA, which necessitates the pre-culture of nematode populations. However, it is efficient in resolving nematode isolates (Ijaz & Khan, 2009).

The RAPD (Random Amplified Polymorphic DNA) method is a novel genetic marker in which specific DNA is amplified using PCR with a primer that is a random sequence of 9–10 nucleotides. However, the drawback is that a single base variation between the complementary oligonucleotide primer and the DNA sequence can result in polymorphism (Liu, 2007).

Among the earliest non-morphological-based techniques for identifying nematodes was the use of enzyme phenotypes (isozyme analyses). Briefly, this method involves removing soluble proteins from whole nematodes in buffer solutions, resolving the extracts in polyacrylamide gel electrophoresis or starch, and then staining for enzymes. Multi-locus enzyme electrophoresis (MEE) is another name for this electrophoretic technique (Esbenshade & Triantaphyllou, 1985). This method clarified evolutionary links, particularly among the major species in the genus *Meloidogyne*, and complemented morphological methods. The process was still laborious and time-consuming, and one of its drawbacks was the requirement to use known samples for reference (Esbenshade & Triantaphyllou, 1990). An overview of different methods is given in Figure 9.1.

Two-dimensional gel electrophoresis (2-DGE) has been used in nematode taxonomy research. The method employs mass-based resolution in a dimension perpendicular to the first dimension, following isoelectric focusing in one dimension to resolve complex protein mixtures by charge. The resolution pattern was then compared among the isolates to identify differences or similarities. Navas and his colleagues used 2-DGE to demonstrate proteomic differences between 18 root-knot nematode isolates representing four species. Some of these changes proved to be species-specific, while others revealed evolutionary links between the different species. An advantage of 2-DGE is that it allows evolutionary inferences to be made about the taxa under study. Mass spectrometry is another method for removing and studying species-specific polypeptides. Disadvantages include the fact that the amount of polypeptides resolved and the polymorphism detected vary depending on the method and sample size (Navas et al., 2002).

A combination of advanced and traditional techniques (morphological, PCR, and protein-based) is helping nematologists identify and manage nematodes. In the future, emerging technologies, such as microarrays, artificial intelligence, autofluorescence, and microfluidic PCR, will help identify problems in a timely manner and protect the world's crops against devastating nematodes.

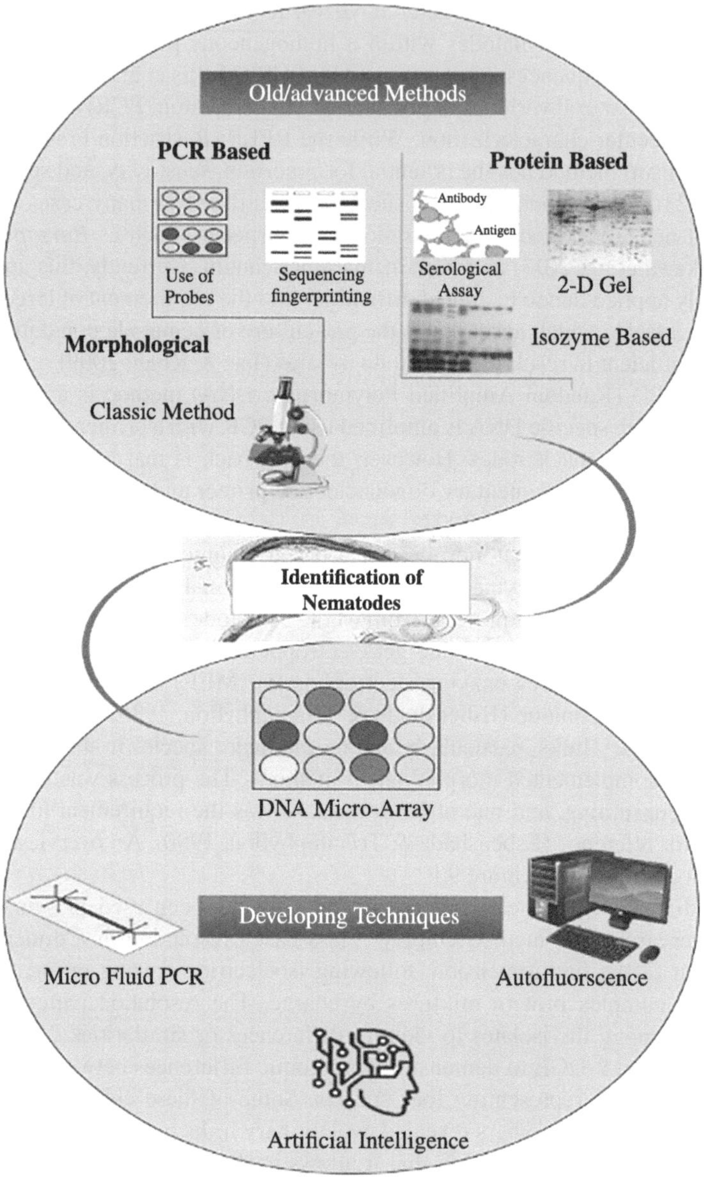

FIGURE 9.1 The comparison between old/advanced and emerging technologies in the field of plant nematology. This figure captures the journey of nematode identification, bridging decades of hands-on science with today's tech-driven breakthroughs. On one side, the trusted tools of the past—microscopes, gels, and painstaking PCR—tell stories of patience and precision. On the other hand, microfluidic chips hum with promise, AI algorithms decode complexity, and autofluorescence illuminates hidden details. Together, they paint a future where farmers and scientists alike can tackle invisible threats faster, smarter, and with fewer resources. It is not just about methods; it is about empowering people to protect crops and ecosystems in a changing world.

Polymerase chain reaction (PCR) has emerged as a crucial procedure in genetic study, which has been a significant part of contemporary biology and biochemistry research. However, because typical 96-well PCR machines have a heat cycling rate of 1–2 °C/s, a full PCR amplification takes 1–2 hours or more. Furthermore, this method necessitates the execution of both sample preparation and post-PCR product detection, which lengthens the analytical process and increases the possibility of cross-contamination. High-throughput nucleic acid amplification is challenging for commercial capillary-based PCR equipment, despite their ability to conduct quantitative analysis by examining melting curves and reaching rapid thermal cycling rates of up to 10–20 °C/s (Walsh et al., 2005). Currently, PCR microfluid chips are preferred over a conventionally sized thermocycler. The advantages of these microchips include lower sample and reagent consumption, quicker analysis times, increased sensitivity, portability, and disposability (Feng et al., 2009).

PCR chips are now highly integrated systems rather than just basic microfluidic parts. A single chip can include several microfluidic processes, including sample preparation, DNA/RNA amplification, and/or product detection. Higher levels of integration can be achieved with micropumps, microvalves, and/or micromixers, particularly dense pumping and valving arrays, which enable the parallel operation of numerous samples to shorten analysis time and minimize or prevent the risks of cross-contamination (Zhang et al., 2007). DNA molecules, such as oligonucleotides or complementary DNA (cDNA), synthesized on silicon chips or glass slides, form microarrays. The microarray DNA probes are hybridized with fluorescently labeled DNA, cDNA, or mRNA extracted from an organism's cells or tissues. A confocal scanner is used to view and scan fluorescent spots for quantitative analysis (Monis et al., 2005). This method combines hybridization with numerous specific oligonucleotide probes and DNA amplification. Many genetic traits can be identified in a single procedure. This method exhibits a high degree of sensitivity and specificity in identifying organisms, while also possessing high throughput capacity. It is a valuable approach to identify one or more parasites that are likely to be present in any ecosystem (Dey & Singh, 2009).

Nematode identification and quantification via image analysis have become possible due to advancements in machine learning, often referred to as deep learning or artificial intelligence (AI). This approach excels at detecting subtle and minute objects, such as nematode eggs, in complex backgrounds and handling large numbers of samples (Yuan et al., 2023). There are various steps involved in machine learning for automated phenotypic detection. To minimize subjectivity, a team of experts first takes many pictures (of nematodes, their eggs, or cysts) and independently labels them. These are then used to construct an algorithm that masks (rejects) the background noise while learning (captures) the important attributes of the objects from the photos in a layer-wise hierarchy. A network model with a supervised learning technique is then used to rebuild the pattern of interest in the input photos (Bogale et al., 2020). To account for differences in background noise across data from various sources, Akintayo et al. (2018) developed a unique end-to-end convolutional selective autoencoder (CSAE) to recognize soybean cyst nematode (SCN) eggs in multiple backgrounds.

9.3 DNA BARCODING

A DNA barcode refers to any DNA sequence used to identify species at any taxonomic level by comparing it with a universal barcode, which is made up of a series of vertical bars printed on commercial goods. Precisely, a DNA barcode is one or more gene sequences extracted from a defined region of the genome to use sequence divergence based on nucleotide alignment to help identify and discover species. Therefore, regardless of the sample's visual classification, the primary purpose of this genetic tool is to efficiently and effectively classify any biological sample by its species, comparing barcode sequences to reference databases (Cold Spring Harbor Laboratory DNA Learning Center, 2020).

The idea was originally put forward by Hebert et al. (2003) as a means of rapid and effective species recognition throughout the tree of life. The underlying principle behind DNA barcoding is that a particular part of the genome, commonly referred to as the 'barcode region,' exhibits sufficient variation to discriminate between species without being too highly conserved within species. This enables the discrimination of closely related taxa and the identification of unknown specimens by comparing their DNA sequences with a reference database.

The first and, so far, most widely used genetic marker for animal barcoding is a standard fragment of approximately 648 base pairs (bp) at the 5' end of the mitochondrial gene coding for the cytochrome c oxidase subunit 1 (COI), a component of an enzyme complex required for oxidative phosphorylation. It has been stated that the COI barcode, which uses universal primers for PCR amplification, may successfully identify species in test assemblages of many animal taxa, primarily insects, birds, fish, and nematodes. These PCR primers are essential to the barcode field and are still widely used today. They were first described for a variety of metazoan invertebrates and produce functional sequences for phylogenetic analysis at the species and higher taxonomic levels (Wilson, 2003).

In addition to COI, other genomic regions, sequences such as the internal transcribed spacer (ITS) of ribosomal DNA, the 18S rRNA gene, and the 28S rRNA gene have been used as barcodes for nematodes. Each of these regions has advantages and limitations, and the choice of barcode depends on the specific research question and the taxonomic group under study (Charrier et al., 2024)

A genomic region must meet three requirements to be used as a DNA barcode: (i) have conserved flanking regions to create universal PCR primers for broad taxonomic application, (ii) have substantial species-level genetic variability and divergence, and (iii) have a short sequence length to support current DNA extraction and amplification capabilities (Kress & Erickson, 2008).

9.4 CHOICE OF BARCODE REGIONS

Selecting an appropriate barcode region is crucial for the success of DNA barcoding. The mitochondrial genome and the ribosomal RNA array are the two genomic sections of the nematode species that have been regularly described. Using the mitochondrial genome as a species-specific diagnostic marker seems logical. The entire circular genome has a large copy number and is composed of conserved regions and rapidly evolving genes, enabling its wide phylogenetic application. The characteristics include conserved structure, gene order, and uniparental inheritance. Gene composition enables ecological and systematic studies that cover the entire animal kingdom. It is interesting to note that the mitochondrial genome of nematodes has unusual structural components, which provide promising diagnostic possibilities. Less repetition of DNA sequence and the same gene order are the basic characteristics of the mitochondrial genome in animals (Brown, 2002). Nematodes, especially the classes Enoplea and Adenophorea, exhibit a distinct variation in gene order. Both major classes, Secernentea (Chromadorea) and Enoplea, are characterized by surprisingly repeated sequence and gene structure. In *Meloidogyne* (the most economically important species of nematodes), repeated regions are converted into species-specific markers, and these markers are now used as a sample to examine other nematode populations (Blok et al., 2002).

For animals (including nematodes), the mitochondrial cytochrome c oxidase subunit I (COI) gene has been widely adopted as the standard barcode region due to its high interspecific variability and ease of amplification (Hebert et al., 2003). The COI gene has proven effective for species identification in a wide range of taxa, including insects, fish, and nematodes (Blaxter et al., 1998; Floyd et al., 2002). The preferred barcodes in animals are mitochondrial DNA (mtDNA) regions over nuclear genomes, even though a mitochondrial barcode is unlikely to indicate nuclear divergence due to the distinct evolutionary histories of nuclear and mitochondrial DNA.

This is because most eukaryotic organisms have mtDNA, which is circular DNA packaged into nucleoids without the protection of histone proteins and is known to be inherited uniparentally from the maternal parent. Despite infrequent recombination, the mitochondrial genome lacks adequate DNA repair mechanisms in comparison to nuclear DNA. This results in a tenfold increase in nucleotide substitution when reactive oxygen species are produced during the respiratory chain. Because mtDNA sequences evolve rapidly, differences between closely related species that have only been separated for short periods may accumulate (Rubinoff, 2006). The 18S rRNA and 28S rRNA genes have also been used as barcode regions, particularly for deep phylogenetic studies, due to their slower evolutionary rates and higher conservation (Holterman et al., 2006). Customarily, the three ribosomal genes (rRNA) 18S, 5.8S, and 28S (26S) are typically included in the rRNA array. These genes have a single unit of 7429–7431 nucleotides long (52a), which are linked in repeating units with their spacer sections in *C. briggsae*. Whereas *C. elegans* is thought to have 100–150 copies of the repeating unit on chromosome 1, where it has 55 tandemly aligned units. Concurrently, there is variation in the composition of the ribosomal array in the various genera of PPNs. The nematodes have a relatively smaller chromosomal location of rDNA. However, variation in the composition of the ribosomal array occurs in the genera of PPNs, that is, *Meloidogyne* (Vahidi et al., 1988). The genus *Meloidogyne* shares a single copy of the 5S ribosomal gene, located in the intergenic spacer region between the 28S and 18S genes. This has been utilized in numerous PCR-based diagnostic steps. *C. elegans* possesses 5S genes at a distance of about 100 copies on chromosome V in a tandem array (Blok et al., 1997).

Besides COI and the ribosomal RNA array, genomic loci other than these have also been tested for DNA barcoding in nematodes. One of the most employed barcode loci for PPNs is the internal transcribed spacer (ITS) region of ribosomal DNA, comprising ITS1 and ITS2. The ITS region is extremely diverse and has been effective in discriminating among species of closely related cyst nematodes (Subbotin et al., 2006) and root-knot nematodes (Adam et al., 2007).

9.5 HOW TO GET A BARCODE?

The primary objective of DNA barcoding is to utilize simple molecular tools to identify unique individuals among their conspecifics, rather than relying on morphology-based techniques, which are time-consuming procedures that require expert taxonomists. The fundamental principle of modern barcodes is based on the observation that, when aligned, some DNA fragments differ only slightly within species while interspecies variation is significantly higher. This difference is known as the 'barcoding gap' (Liu et al., 2011). The selection of the short DNA fragment thus plays a significant role in determining the targeted species among the large number of samples. Slowly evolving gene regions frequently do not differ across closely related organisms. In contrast, rapidly evolving DNA sequences can rewrite the remnants of old affinities, adding more sequence variety and enhancing the ability to differentiate between species. Furthermore, the selected DNA segment needs to be standardized, accessible across different taxonomic groupings, and have conservative primer binding sites for the barcode marker to be technically amplified and sequenced from a small amount of specimen using the polymerase chain reaction (PCR) (Steinke et al., 2016).

However, it is unlikely that there will be a single universal DNA barcode for all kinds of life, despite tremendous efforts to identify a single section of DNA that satisfies all the above-mentioned criteria. The primary reason for this is that barcoding zones are suitable in some species but not in others, as they have not evolved neutrally since the period of speciation and are often influenced by weak positive or negative selection. As a result, multi-locus barcodes targeting several extant taxa have been created and evaluated for their ease of amplification and ability to resolve species as part of the barcode validation procedure.

DNA barcoding offers several advantages over traditional morphological identification methods. It provides a rapid and cost-effective means of species identification, particularly for taxa that are cryptic in nature and are difficult to distinguish based on morphological characteristics alone. For example, many species of PPNs exhibit overlapping morphological traits, making their identification challenging even for experienced taxonomists (De Ley et al., 2005). DNA barcoding overcomes this limitation by providing a molecular signature that is unique to each species. Nematode community surveys highlight the challenges of species identification. A substantial number of keys demand crucial choices regarding adult traits for precise identification. Juveniles make up a significant portion of nematode populations. DNA barcoding enables the identification of all life stages of nematodes, including eggs, juveniles, and adults, which may not be morphologically distinguishable (Shao et al., 2023).

In nematode groups like *Dorylaimida*, both sexes are necessary for successful key navigation. However, specimens are so rare that species identification must be based on a single sex. In specific cases, all the life stages can be identified using molecular markers. For example, the identification of infectious juveniles of *M. mayaguensis* in soil has only been accomplished using molecular techniques (Brito et al., 2004). Different life stages may have varying impacts on plant health and require different management strategies (Ali et al., 2015). For that reason, molecular identification is crucial for PPNs (Powers, 2004). DNA barcoding has discovered cryptic species that are morphologically similar but genetically distinct. For example, DNA barcoding has revealed the existence of cryptic species within the *Meloidogyne* genus, which were previously considered a single species (Tigano et al., 2005).

Predicta B is a DNA-based soil testing service for arable farming. It helps cereal and pulse growers identify which pathogens pose a risk to their crops before planting. A national survey of root lesion nematodes conducted by DAS00137 and DAV00128 confirmed that PREDICTA B tests are reliable for *P. neglectus*, *P. thornei*, *P. quasitereoides*, and *P. penetrans*. Therefore, PREDICTA B can be used with confidence to determine which of the major root lesion nematodes are present in a field. Nematode species are present in a paddock and the level of each species prior to planting. The research has confirmed that *P. thornei* is generally the main species on the higher clay soils in the black/grey soils in the north, while *P. neglectus* tends to be the dominant species in central New South Wales (NSW) in Australia and further south on lower clay red/brown soils (McKay et al., 2018). This has important implications for nematode taxonomy and the development of species-specific control measures. However, modern identification techniques must be accurate, quick, dependable, affordable, and, most importantly, should allow for the characterization of scientifically novel specimens. Genomics has sped up the identification of new genes and the comparison of gene functions. DNA barcoding will enable users to quickly identify recognized species and hasten the discovery of species that have not yet been discovered in nature.

Nematodes typically have 1000–2000 cells and measure around 1 mm in size (Powers, 2004). This size requires little work on the part of the sample preparation and is conveniently appropriate for a PCR reaction. A nematode has no complex structural components that need to be disassembled or manipulated to access its DNA. The nematode can frequently be crushed in a droplet of water, and then some of the resulting solution can be added straight to a PCR experiment. Additional preparatory techniques include DNA purification and proteinase digestion. A single nematode can frequently produce several amplifications. For later study, the fluid from crushed nematodes can be used as a DNA voucher and stored frozen indefinitely (Powers & Harris, 1993). Advanced technologies and biocides are no longer considered viable alternatives to managing nematodes. These approaches must be able to control the nematodes by focusing on certain species, species groups, or taxa. Future diagnostic techniques will need to be more specific for a wide variety of nematode taxa (Barker & Koenning, 1998).

A single species of PPN rarely feeds on a single plant in an agroecosystem. In many cases, the community composition of plant-feeding nematodes in an agronomic soil reflects the order of cropping systems and rotation schemes across several years. Usually, multiple plant-parasitic species are recovered from the roots of a single plant. It has been suggested that a crucial element of the contemporary concept of integrated pest control is the monitoring of the identification and abundance of every nematode in the soil ecosystem. According to researchers, nematode diversity is higher than previously thought, even when a subset of the nematode community is targeted for control. For instance, it appears that a variety of root-knot species (*Meloidogyne*) are present in numerous agronomic soils, and each species has distinct host ranges and life-history traits. This can also be identified by DNA barcoding, which could make management easier (Neher, 2001).

9.6 STANDARDIZATION AND QUALITY CONTROL

Standardization of protocols and quality control measures are essential to ensure the reliability and reproducibility of DNA barcoding results. The Consortium for the Barcode of Life (CBOL) has made recommendations on DNA barcoding standards, including barcode region choice, primer design, and sequence quality requirements (Ratnasingham & Hebert, 2007). These standards serve to reduce errors and allow DNA barcoding data to be compared among studies. Quality control involves the application of positive and negative controls in PCR amplification, confirmation of sequence quality by chromatograms, and alignment of sequences against reference databases to validate species identification (Hajibabaei et al., 2007). Furthermore, the application of more than one barcode region can enhance accuracy in species identification, especially for taxa with high levels of genetic diversity.

9.7 METHODOLOGIES FOR DNA BARCODING OF PPNS

DNA barcoding of the PPNs uses standardized methods to guarantee reproducibility and accuracy. Below are key methodologies supported by references from the literature.

9.8 SAMPLE COLLECTION AND PRESERVATION

Since most migratory PPNs are located near plant roots, rhizosphere samples are ideal. More nematodes may be found in samples taken from nearby, less impacted plants because severely stunted plants may not have an extensive enough root system to maintain many of them. Typically, soil samples are collected 15–20 cm below the surface, but nematodes that damage tree crops and other perennial plants with deep roots can be found 60 cm deep from the ground soil. It is best to keep plant material and soil samples moist for the inspection of nematodes. Whole plants should be stored separately from soil, although polythene bags make great sample containers for soil and/or roots. Therefore, samples should be stored at a temperature of 5°C in temperate regions, 10–16°C in intermediate latitudes (with warm climatic conditions), and 16–18°C in tropical and subtropical regions. While storing samples in refrigerators is a standard procedure, the ability of some nematodes to recover from tropical soils can be negatively impacted by low temperatures (-5°C) (Whyte & Gowen, 1974).

Formalin–acetic acid (4:1) or 5% formalin (2% formaldehyde solution) can be added to roots and shoot tissues to fix them for storage and further analysis, such as staining. Another option is to immediately immerse fresh material in hot lacto-phenol/lactoglycerol, which softens tissues and is very beneficial for *Meloidogyne* females recovering from roots. Elmiligy and Grisse (1970) combined soil samples with a hot fixative (100 mL of 40% formaldehyde, 10 mL of glycerol, and 890 mL of distilled water at about 80°C). Centrifugal flotation is used to remove nematodes from soils that have undergone fixation treatment.

9.9 NEMATODE EXTRACTION

PPNs are typically extracted from soil or plant tissues using techniques such as Baermann's funnel, centrifugal flotation, or sieving (Janssen et al., 2017). Basic supplies needed for the extraction and management of PPNs can be purchased at the neighborhood market (such as sieves, dishes, flasks, filters, funnels, and tubing) or built in person (such as nematode transfer picks, counting dishes, sieves, and racks). For nematode extraction, plastic or stainless steel is preferred over brass or bronze gauze, rings, or pans because metallic ions, particularly copper, can be hazardous to nematodes (particularly Dorylaimia) when released into small amounts of static water (Pitcher & Flegg, 1968). However, it does not seem hazardous to come into short-term contact with metal sieves, as in the screening procedure.

9.10 DNA EXTRACTION

DNA is extracted from single nematodes or pooled samples using the following methods.

9.11 CTAB-BASED METHODS

Nematode samples are placed in 200 μL CTAB buffer and 2 μL mercaptoethanol and incubated for 30 min at 65°C, with 10 s rotations every 10 min. After centrifugation for 5 minutes at 2400 relative centrifugal force (RCF), the samples are centrifuged for 10 minutes at 9600 RCF at room temperature after the addition of chloroform:isoamyl alcohol (CIAA) in a ratio of 24:1 and homogenization for 20 seconds. Using a micropipette, the top layer is transferred to a fresh tube, while the pellet (bottom layer) remains in the original tube. Cooled 96% absolute ethanol (two times the volume of the sample) is then added to homogenize the mixture. The homogenized samples are incubated at -20°C. Centrifuge at 9600 RCF for 15 minutes at room temperature. The upper layer is gradually removed to collect the pellets. Chilled full-volume 70% ethanol is then added and stirred until evenly distributed, and the mixture is centrifuged at 9600 RCF for 15 minutes. The remaining pellets are then discarded when the ethanol solution is removed. Eppendorf tubes filled with DNA pellets are placed either in a laminar airflow or on an inverted tissue to dry for three hours. The DNA pellets are then placed in 30 μL of TE buffer (10 mM Tris-Cl, 0.5 mM EDTA at pH 9.0) and stored at -20°C. This method is used for high-quality genomic DNA (Holterman et al., 2006).

9.12 COMMERCIAL KITS

Various commercial kits, such as the DNeasy Blood & Tissue Kit (Qiagen), are optimized for extracting DNA from nematodes (Lombard et al., 2024). The Genomic DNA Mini Kit (Tissue Kit) protocol from the Geneaid Taiwan manual describes the extraction of DNA as follows: After crushing the nematode-infected root sample, 200 µL of tissue dissociation buffer and 20 µL of proteinase K are added to the tube and shaken until the mixture becomes homogenized. The suspension is then incubated for 30 minutes at 60°C. The tube is rotated every five minutes during the incubation period. Next, 200 µL of cell lysis buffer is added, and the mixture is stirred for five seconds. To ensure the lysate is clear, the suspension is incubated again for 20 minutes at 60°C. Following that, 200 µL of cold 100% ethanol is added and mixed thoroughly. After transferring into a 2 µL collecting tube and centrifuging at 15,000 RCF for two minutes, the suspension, including the precipitate, is moved to a GS column (spin column tube supplied by the kit). The GS column is then moved to a fresh 2 mL tube. After adding 400 µL of protein removal buffer to the GS column, it is centrifuged for 60 seconds at 15,000 RCF. The GS column is placed into the same 2 mL tube after the suspension is disposed of. To remove the salt residues, 600 µL of washing buffer is then added to the GS column and again centrifuged for 60 seconds at 15,000 RCF. To dry the column matrix, the suspension is disposed of one more time and centrifuged for three minutes at 15,000 RCF. A fresh 1.5 mL tube is then filled with the GS column. After inserting the heated 30 µL elution buffer into the middle of the column matrix, it is left to stand for at least five minutes. After that, it is centrifuged again for 60 seconds at 15,000 RCF. The suspension in the 1.5 mL tube is prepared for storage at -20°C or for further use (Geneaid Biotech Ltd., 2017).

9.13 CHELEX EXTRACTION

A chelating resin with a high affinity for polyvalent metal ions is called Chelex 100. The copolymers of styrene divinylbenzene that make up the Chelex 100 resin have paired iminodiacetate ions that serve as chelating groups. The presence of Chelex 100 during boiling is thought to inhibit DNA degradation by chelating metal ions that could catalyze the destruction of DNA exposed to high temperatures in solutions with low ionic strength. In the basic Chelex 100 process, the sample is boiled in a 20% (w/v) Chelex 100 solution, and then some of the supernatant is added straight to the PCR mix. Sample DNA is denatured as a result of this process (Forensic and Scientific Services, 2007). This method is used for rapid PCR-ready DNA, though with lower purity.

9.14 TARGET GENETIC MARKERS

9.14.1 S rRNA

The use of 18S rRNA as a target genetic marker helps identify nematodes at the molecular level. This genetic marker is used for broad phylogenetic analysis

(Blaxter et al., 1998). The amplicon 18S small subunit ribosomal RNA (SSU) primers capture more heterogonous species than other taxa-specific primers; therefore, they are suitable for DNA metabarcoding. Consequently, Kenmotsu et al. (2020) used this method to perform a taxonomic analysis of soil nematode colonies. Nematodes were categorized using these sequences to create operational taxonomic units (OTUs), which are taxonomic groups that share identical DNA barcode sequences. In each OTU, the number of nematodes indicates the percentage of each taxonomic group in the nematode community. In contrast, the number of OTUs reflects the number of taxonomic groups (i.e., nematode species diversity). This approach has also been used to analyze the taxonomy of soil nematodes. High-throughput amplicon sequencing using Illumina MiSeq technology was used to resolve the taxonomy of 68 out of 96 individual nematodes isolated from forest soil to determine which of the four SSU gene regions (regions 1–4, 337–388 base pairs (bp) in length) were acceptable for this procedure (Kenmotsu et al., 2020).

9.15 INTERNAL TRANSCRIBED SPACER

Most eukaryotes have one or more blocks of tandemly organized rDNA repeats. Areas transcribed into non-conserved parts of the pre-rRNA are known as transcribed spacers, which are further classified into external (ETS) and internal (ITS) regions. Transcription units alternate with spacer regions, which are often referred to as non-transcribed spacers (NTS) (Long & Dawid, 1980). Moreover, identification at the species level is facilitated by the ITS region, which is a highly variable and adaptable genetic marker (Subbotin et al., 2006) situated between the 18S and 28S ribosomal DNA genes (Saeki et al., 2003). The internal transcribed spacer (ITS) contains most of the sequence diversity in the rDNA. The 5.8S coding region in the rDNA cistron interrupts the ITS into ITS1 and ITS2, making the ITS helpful for molecular systematics of closely related nematode species (Bogale et al., 2020).

9.16 CYTOCHROME C OXIDASE I

Most research favors the mitochondrial cytochrome c oxidase subunit I (COX1) and rDNA genes due to the variable regions that conserved ones surround. While COX1 is better suited for resolution at lower taxonomic levels, such as species and subspecies groups, due to the higher level of sequence diversity in the variable region, rDNA is more appropriate for use at wider taxonomic levels due to the higher level of sequence conservation in the flanking regions, which enables the design of 'universal' primers. Another benefit of COX1 and rDNA is that they are found in numerous copies in nematode genomes, which makes it possible to perform PCR amplifications using small amounts of DNA templates, particularly in single nematodes. The generated sequence information is then utilized to resolve and/or identify the taxa involved in character-based or phylogenetic analyses, the latter of which permits evolutionary inferences (Donn et al., 2011). The COI is a standard animal barcode that requires nematode-specific primers (Porazinska et al., 2009).

9.17 PCR AMPLIFICATION AND SEQUENCING

9.17.1 Primers

Species-specific primers are made using parts of the nematode genome that differ from one species to another, such as the ITS region of ribosomal DNA. It is possible to choose the 18S-ITS1 rDNA region as the amplification target because the universal forward primer targets the 18S region, whereas species-specific reverse primers target the ITS1 region (Huang et al., 2019). Primers are less expensive, less time-consuming, and less technically challenging to utilize, and hence are more suitable for daily and frequent diagnostic applications. Today, end-point single PCR tests using species-specific primers are used in diagnosis. The strength of the multiplex PCR assay is that it can screen a large number of gene targets in one reaction using minimal DNA template and reagents to carry out a diagnostic (Boutsika et al., 2004). PPN diagnosis can also be performed using some alternative primers (e.g., universal primers, such as SSU_18A/SSU_26R to diagnose 18S rDNA, or nematode-specific primers) (Derycke et al., 2005; Vrain et al., 1992).

9.17.2 Nested PCR

The target DNA sequence is amplified by two sets of primers, making nested PCR more sensitive compared to single-round PCR. Employing two sets of primers enhances the specificity of the reaction and minimizes the chance of amplifying non-target DNA. The second set of primers will not bind if the first PCR amplifies the wrong target since further amplification will not be possible (Futai et al., 2005). Nested PCR is more precise for the identification of particular nematode species compared to parasitological methods. It can also be adjusted by incorporating new primers within the second PCR while reusing the outer primer of the first PCR. This is referred to as semi-nested PCR (Trasia, 2020).

9.18 SANGER SEQUENCING

The options provided by Sanger sequencing can be used to accomplish the identification of PPN species on a morphological basis. Sanger sequencing remains the gold standard procedure in terms of quality. It is the best way to precisely identify small insertions and deletions, such as single nucleotide variations, compared to massively parallel sequencing. Sanger sequencing is less rigid in the detection of a particular species. It saves cost when a particular sample requires urgent testing and cannot be scheduled conveniently. Compared to other sequencing techniques, Sanger sequencing can easily cover longer segments, such as 1000 base pairs, unless other superior techniques are in use (van de Vossenberg et al., 2023).

9.19 SEQUENCE COMPARISON AND DATABASE ALIGNMENT

Nematode DNA sequences are often compared against databases such as GenBank or BOLD (Barcode of Life Data Systems) once sequenced. The significance of these nematodes lies in the extreme importance of databases in the identification, measurement of diversity, and ecology of nematode populations. The taxonomic placement of nematodes can be estimated by comparing their sequences to databases, which help determine the genus, family, and species of the nematodes. These databases are more accurate in their taxonomic assignments than other databases. For example, 18S NemaBase has increased diversity scores and taxonomic identifications.

Integrated analyses can be performed using genomic, transcriptomic, protein, and functional annotation data listed in databases such as the Plant Parasitic Nematodes Database (PPND). Using Internet-based databases, such as NemaTaxa, scientists can explore the ecology of nematode communities, detect taxa at different levels, and learn more about soil ecosystems. Locust or nematode molecular characterization and prediction of nematode diversity require reference libraries such as GenBank, nematode databases, and BOLD. These comparisons guide morphological determination to the species of PPNs. BOLD and GenBank databases are invaluable and straightforward in the curation process of DNA barcoding. Delimitation (e.g., >97% ITS similarity threshold between conspecifics) is also used in comparison with species thresholds (Elena & Lalić, 2013).

9.20 EMERGING TECHNOLOGIES

9.20.1 Next-Generation Sequencing (NGS)

Next-generation sequencing (NGS) facilitates the metabarcoding of nematode communities. Increasing use of NGS and metabarcoding for the analysis of nematode communities has been observed (Treonis et al., 2018).

Nuclear ribosomal RNA gene repeats (rDNA) are the most widely used markers for nematode metabarcoding. Amplification of the D3 expansion region of the 28S rDNA has proven to characterize around 90% of the species sampled (Porazinska et al., 2009).

9.21 LOOP-MEDIATED ISOTHERMAL AMPLIFICATION

A sensitive, specific, and rapid loop-mediated isothermal amplification (LAMP) assay for nematode detection has been developed for quarantine, border inspection, and field surveillance. The assay is based on the variable region sequence of the nematode mitochondrial cytochrome oxidase subunit I (COI) gene. The method is very sensitive as it can detect a single nematode crude DNA sample or 10^3 copies/μL of cloned target plasmid. Results of detection can be easily ascertained by a color change of the hydroxynaphthol blue dye or an indicator bar on a particular lateral flow dipstick. The assay takes a minimum of 35 minutes to complete. A mitochondrial gene has been used as a marker to distinguish *A. besseyi* rice race from closely related *A. besseyi* nest fern pathotypes. In nurseries, this assay is ideal for seedling screening and phytosanitary inspection (Yang & Yu, 2019).

9.22 DNA BARCODING IN BASIC STEPS

In short, the basic procedure of DNA barcoding of PPNs is described as follows:

- Identification of symptoms of infected plant
- Collection of samples
- Extraction of Nematodes (e.g., Bearman's funnel method)
- Extraction of DNA from isolated PPNs
- Selection of genetic marker
- Performing PCR
- Comparing the resulted PCR product with different databases (e.g., BOLD)

A visual representation of nematode DNA barcoding identification—from sample collection to comparison with databases—is given in Figure 9.2 (Tileubayeva et al., 2021).

9.23 APPLICATIONS OF DNA BARCODING IN PPN RESEARCH

DNA barcoding has emerged as a transformative tool in plant-parasitic nematode (PPN) research, enabling accurate species identification, ecological insights, and innovative management strategies. Accurate identification of PPNs is critical for disease management because many species share overlapping morphological traits. DNA barcoding resolves taxonomic ambiguities by targeting conserved and variable genetic regions.

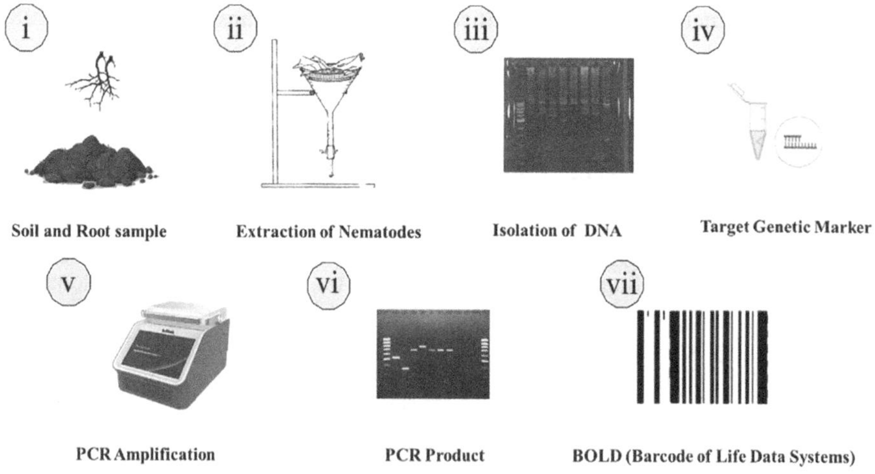

FIGURE 9.2 (i) Root or soil sample collection, (ii) extraction of nematodes by Baermann's funnel method, (iii) isolation of DNA on gel, (iv) choice of targeted marker gene, (v) PCR amplification, (vi) visualization of PCR product on gel documentation system, and (vii) comparison of results with the Barcode of Life Data System (BOLD).

The mitochondrial COI gene and ribosomal ITS regions distinguish cryptic species such as *Meloidogyne incognita*, *M. javanica*, and *M. arenaria*, which cause identical galling symptoms but differ in host range and virulence. Subbotin et al. (2006) used ITS to clarify the phylogeny of *Tylenchida* nematodes. Species like *Globodera pallida* and *G. rostochiensis* threaten global potato production and are reliably distinguished using ITS rDNA sequences. Lombard et al. (2024) optimized PCR protocols for these species, and Waeyenberge et al. established ITS markers for *Heterodera* identification. DNA barcoding reveals cryptic genetic diversity in morphologically homogeneous species, which helps in pest control and biosecurity. Palomares-Rius and his friends identified cryptic lineages in *Pratylenchus thornei* and *P. neglectus* using 28S rDNA and COI markers (Palomares-Rius et al., 2017). Similarly, Derycke et al. (2005) observed cryptic speciation in marine nematodes, illustrating the wider utility of mitochondrial markers.

Vrain et al. (1992) employed ITS and COI sequencing to discriminate between species within a vector clade that carries plant viruses. Ye et al. (2015) clarified phylogenetic relationships in *Aphelenchoides* using COI. Environmental DNA (eDNA) metabarcoding allows high-throughput profiling of nematode assemblages, connecting PPN diversity to ecosystem. Kageyama and Toju (2022) applied metabarcoding with 18S and ITS primers to investigate nematode diversity in agricultural soils. Van Den Hoogen et al. (2019) performed a global soil survey to uncover associations between PPN abundance and soil characteristics.

DNA barcoding clarifies evolutionary history and invasion routes of PPNs. Blaxter et al. (1998) developed a molecular framework of Nematoda with SSU rDNA. Holterman et al. (2006) extended this work to clarify deep phylogenetic nodes in Tylenchida. Brito et al. (2004) tracked the worldwide distribution of the invasive *Meloidogyne enterolobii* with COI haplotypes, guiding quarantine practices. DNA barcoding facilitates international phytosanitary regulations through timely detection of regulated pests. The European Plant Protection Organization (EPPO) requires DNA barcoding for identification of quarantine nematodes. Niu with his colleagues developed a LAMP assay targeting COI for on-site detection of *Meloidogyne* spp., minimizing dependence on laboratory facilities (Niu et al., 2011).

Knowledge of PPN virulence and host specificity is useful in resistance breeding and sustainable agriculture. Williamson et al. applied PCR-based markers to diagnose *Meloidogyne chitwoodi* populations pathogenic to tomato cultivars. Subbotin et al. (2023) employed metabarcoding to analyze PPN–host plant–soil microbiome interactions and discovered niche partitioning among *Heterodera* communities.

9.24 CHALLENGES

Although DNA barcoding has numerous benefits, it is not without limitations. One of its main challenges lies in the existence of intraspecific variation, which may overlap with interspecific variation, causing difficulties in species discrimination. This is especially problematic for species that are genetically diverse or rapidly speciating (Meyer & Paulay, 2005). The efficacy of DNA barcoding relies on the availability of reference sequences in public databases. For many nematode species, such sequences are incomplete or non-existent (Workentine et al., 2020). Reference sequences in public databases such as GenBank and the Barcode of Life Data System (BOLD) are often unavailable. Identification at the species level depends on the presence of high-quality reference sequences, which are typically absent for most nematode species, particularly those from understudied regions (Porazinska et al., 2009). Porazinska et al. (2009) also reported that only ~20% of described PPN species had COI barcodes in BOLD up to 2009, a percentage still below optimal today (Holterman et al., 2006). Erroneous entries in databases can result in species misclassification. For example, sequences reported as *Heterodera glycines* were later shown to represent undetected species (Subbotin et al., 2006).

No universal barcode region is optimal for all nematode taxa, and marker selection has a significant impact on resolution and reliability. While COI is widely used in animals, it may not be suitable for all nematode taxa due to variations in amplification success and sequence quality (Blaxter et al., 1998), and the choice of barcode region can influence the success of DNA barcoding. The degree of universality of COI is also questioned for several reasons. For example, its applicability to a wider range of animal species is hampered by the considerable diversity of nucleotide sequences at COI priming sites. To address this issue, cocktails of degenerate primer sets have been used for barcoding purposes in different animals.

As COI is the standard animal barcode, its utility in nematodes is inconsistent Blaxter et al. (1998) observed that COI amplification in about 30% of nematode species is unsuccessful because of primer mismatches or lack of variation. For instance, *Xiphinema americanum* has very low COI diversity even though there are a number of cryptic lineages (Vrain et al., 1992). Degraded DNA from stored specimens or from environmental samples lowers the success with which PCR is doneperformed. Certain barcode regions may be hard to amplify because of primer mismatches or because the sample contains PCR inhibitors (Porazinska et al., 2009). Soil extract humic acids (HA) inhibit PCR, requiring special extraction purification techniques (Kosaka & Kanzaki, 2019). Ribosomal targets such as ITS and 18S rRNA are commonly utilized as alternatives. Nonetheless, ITS might be non-resolving in taxonomically newly diverged species (e.g., Globodera rostochiensis and G. pallida), whereas 18S is too conserved to resolve species-specific discrimination (Lombard et al., 2024; Subbotin et al., 2006).

Mitochondrial (COI) and nuclear (28S, ITS) combined markers are more accurate but more costly and complicated (Derycke et al., 2005). Although DNA barcoding is normally affordable, start-up equipment and reagent expenses could be too high for some labs. High-throughput sequencing and bioinformatics are expensive to establish and require substantial technical and financial resources, making access difficult in low-income countries (Thompson et al., 2021).

One of the major difficulties of DNA barcoding is resolving intraspecific genetic variation from interspecific divergence. Overlapping of genetic distances among closely related species may create identification problems, especially taxa with high genetic diversity or rapidly speciation. Meyer and Paulay (2005) pointed out that intraspecific mitochondrial COI divergence in certain marine invertebrates coincides with interspecific variation, a trend similarly shown in nematodes. Cryptic species from the same genus, Meloidogyne (e.g., M. incognita and M. javanica), exhibit low COI divergence even though they are morphologically different (Blaxter et al., 1998).

Overlapping makes diagnostics difficult in mixed infestations (Janssen et al., 2017). Lineages such as Pratylenchus spp. exhibit great mitochondrial haplotype diversity within one morphospecies, which obfuscate species boundaries (Elena & Lalić, 2013). In such instances, complementary morphological or multi-locus studies are necessary (Derycke et al., 2005).

9.25 FUTURE DIRECTIONS

The future of DNA barcoding for the study of PPNs is bright, with a number of developments on the horizon that will be exciting. Improvements and enlargements of reference sequences available in public databases are underway and will improve the accuracy and reliability of DNA barcoding. Projects such as the Barcode of Life Data System (BOLD) and international collaboration with global biodiversity initiatives work to bridge sequence gaps in nematode coverage, especially in understudied groups and areas (Nadler, 2002; Ratnasingham & Hebert, 2007).

The combination of DNA barcoding with recent instruments and procedures allows us to diagnose and control parasitic nematodes in plants on time. NGS technologies provide high-throughput sequencing of DNA barcodes, enabling researchers to conveniently analyze sophisticated nematode communities. Illumina platforms (MiSeq, NovaSeq) are commonly employed for metabarcoding because of their precision and scalability. Kageyama and Toju employed Illumina sequencing to represent nematode diversity in agricultural soils with 18S rRNA and ITS markers, identifying rare PPN species in addition to free-living nematodes (Kageyama & Toju, 2022).

Nanopore sequencing offers long-read functionality and portability, enabling in situ field diagnostics. Its applicability for the direct identification of eukaryotic pathogens, including nematodes, from environmental samples is highly convenient. This technology is especially useful for quarantine species like *Globodera pallida* for rapid detection (Sonawala et al., 2024). High-level bioinformatics pipelines are needed to process and interpret large NGS datasets. BLAST (basic local alignment search tool) is still core for sequence comparison against reference databases. Newer versions (e.g., BLAST+) are specifically tuned for NGS data (Sirangelo & Calabrò, 2020). MEGA (Molecular evolutionary genetics analysis) software provides robust construction of phylogenetic trees with maximum likelihood approaches. This software has unraveled the evolutionary relationships between cryptic PPNs complexes like *Meloidogyne* spp. (Elena & Lalić, 2013). Tools like QIIME 2 and DADA2 analyze raw sequencing reads to generate amplicon sequence variants (ASVs), thereby enhancing taxonomic resolution. These pipelines have been used to study PPN communities in agricultural soils (Kosaka & Kanzaki, 2019).

New genomic regions are being investigated by researchers that could potentially provide higher resolution for discriminating between species, especially in taxonomically difficult groups. For instance, mitochondrial genomes, whole-genome skimming, and ultra-conserved elements (UCEs) are being evaluated for their efficiency in untangling nematode phylogenies that are difficult to resolve (Adams et al., 2006; De Ley & Blaxter, 2002; Vancaester & Blaxter, 2023). The integration of DNA barcoding with transcriptomics, proteomics, and metabolomics will yield a more comprehensive understanding of nematode biology and parasite–host interactions. Multi-omics strategies are already revealing effector protein function and host immune evasion strategies (Haegeman et al., 2011; Kikuchi et al., 2011; Zahoor et al., 2023). The advancement of handheld sequencing technologies, such as the Oxford Nanopore MinION, will facilitate field-based real-time DNA barcoding, thereby transforming nematode diagnostics and monitoring. These technologies enable the rapid on-location identification of invasive species, with reduced dependence on laboratory facilities (Boykin et al., 2019; Ip et al., 2015; Menegon et al., 2017). The use of advanced methodologies in the scientific field also has holds bright promise for a better future. NGS-based metabarcoding measures the diversity of PPNs as a function of soil health and agriculture (Van Den Hoogen et al., 2019) combining Illumina sequencing with bioinformatics to investigate worldwide nematode communities and determine PPN hotspots. It aids in tracing community profiling. Handheld sequencers such as the ONT MinION facilitate on-field detection of Meloidogyne spp. in roots, minimizing diagnostic turnaround time (Thompson et al., 2021).

9.26 CONCLUSION

DNA barcoding has proven to be a valuable tool in PPN identification and research. By providing a quick, reliable, and affordable means of discriminating species, DNA barcoding has revolutionized nematode taxonomy, diagnosis, and ecological investigations. Despite its difficulties, the combination of DNA barcoding with new technologies, such as NGS and bioinformatics, has tremendous potential to increase knowledge of nematode biodiversity and enhance the management of these devastating pests. With the progress of science, DNA barcoding will certainly play a crucial role in nematode agriculture and research.

The use of DNA barcoding has come as a revolutionary tool for identifying, classifying, and managing PPNs, and it has successfully addressed fundamental issues presented by the inconspicuous morphology, microscopic size, and use as a host by PPNs. It enables fine species delineation, even between closely related species, through the application of standardized genetic markers such as the COI, ITS, and 18S/28S rRNA regions, and has the advantages of being faster, scalable, and cheaper as a diagnostic method. Use of DNA barcoding with high-throughput sequencing and bioinformatics has also increased its appeal in cryptic diversity identification, invasive species tracking, and nematode-plant interaction sorting. Despite the positive aspects, some aspects have not been fulfilled, such as the need to have strong reference databases, barcoding protocol standardization, and control of intra- and interspecific genetic variation. Collaborative projects to expand genomic repositories, pilot-test multi-locus barcoding schemes, and standardize global taxonomic systems will be essential for the generation of optimal precision and reproducibility. Moreover, the combination of DNA barcoding with complementary tools (e.g., metabarcoding, phylogenomics) will also enhance knowledge of nematode biodiversity and their evolutionary and ecological roles.

To the agricultural sector, DNA barcoding has possible practical benefits, including increasing quarantine measures, resistance breeding programs, and providing knowledge on sustainable pest control practices. With technologies demystifying the power to utilize molecular resources, these technologies should be integrated into field-based diagnostics to give farmers and policymakers a fundamental advantage over crop losses caused by nematodes. Ultimately, the current paleontologically based taxonomic discipline is brought to the convergence of hopes and solutions for achieving food security and ecosystem integrity goals through the keystone of modern nematology—specifically, DNA barcoding.

REFERENCES

Adam, M. A. M., Phillips, M. S., & Blok, V. C. (2007). Molecular diagnostic key for identification of single juveniles of seven common and economically important species of root-knot nematode (*Meloidogyne* spp.). *Plant Pathology*, *56*(1), 190–197.

Adams, B. J., Fodor, A., Koppenhöfer, H. S., Stackebrandt, E., Stock, S. P., & Klein, M. G. (2006). Reprint of "Biodiversity and systematics of nematode-bacterium entomopathogens". *Biological Control*, *38*(1), 4–21.

Aikawa, T., Kanzaki, N., & Maehara, N. (2013). ITS-RFLP pattern of *Bursaphelenchus xylophilus* (Nematoda: *Aphelenchoididae*) does not reflect nematode virulence. *Journal of Forest Research*, *18*, 384–388.

Akintayo, A., Tylka, G. L., Singh, A. K., Ganapathysubramanian, B., Singh, A., & Sarkar, S. (2018). A deep learning framework to discern and count microscopic nematode eggs. *Scientific Reports*, *8*(1), 9145.

Ali, M. A., Abbas, A., Azeem, F., Javed, N., & Bohlmann, H. (2015). Plant-nematode interactions: From genomics to metabolomics. *International Journal of Agriculture and Biology*, *17*(6), 1071–1082.

Antil, S., Abraham, J. S., Sripoorna, S., Maurya, S., Dagar, J., Makhija, S., Bhagat, P., Gupta, R., Sood, U., Lal, R., & Toteja, R. (2023). DNA barcoding, an effective tool for species identification: A review. *Molecular Biology Reports*, *50*(1), 761–775. https://doi.org/10.1007/s11033-022-08015-7

Anwar, H., Jabran, M., Moosa, A., Arshad, U., Haseeb, A., Jabbar, A., Burhan, M., Abbas, A., Naveed, M., & Ali, M. A. (2022). Effect of sole and consortium application of endophytic bacteria on plant growth promotion and inhibition of *Meloidogyne incognita* infection in okra. *Pakistan Journal of Nematology*, *40*(2), 138–146.

Arshad, U., Jabran, M., Ahmed, S., Abbas, A., Jabbar, A., Zahid, M. S., & Ali, M. A. (2022). Seed-priming: A novel approach for improving growth performance and resistance against root-knot nematode (*Meloidogyne incognita*) in bread wheat (*Triticum aestivum* L.). *Gesunde Pflanzen*, 1–11. https://doi.org/10.1007/s10343-022-00672-9

Barker, K. R., & Koenning, S. R. (1998). Developing sustainable systems for nematode management. *Annual Review of Phytopathology*, *36*(1), 165–205. https://doi.org/10.1146/annurev.phyto.36.1.165

Blaxter, M. L., De Ley, P., Garey, J. R., Liu, L. X., Scheldeman, P., Vierstraete, A., Vanfleteren, J. R., Mackey, L. Y., Dorris, M., & Frisse, L. M. (1998). A molecular evolutionary framework for the phylum Nematoda. *Nature*, *392*(6671), 71–75.

Blok, V. C., Phillips, M. S., & Fargette, M. (1997). Comparison of sequences from the ribosomal DNA intergenic region of *Meloidogyne mayaguensis* and other major tropical root-knot nematodes. *Journal of Nematology*, *29*(1), 16.

Blok, V. C., Wishart, J., Fargette, M., Berthier, K., & Phillips, M. (2002). Mitochondrial DNA differences distinguishing *Meloidogyne mayaguensis* from the major species of tropical root-knot nematodes. *Nematology*, *4*(7), 773–781.

Bogale, M., Baniya, A., & DiGennaro, P. (2020). Nematode identification techniques and recent advances. *Plants*, *9*(10), 1260. https://doi.org/10.3390/plants9101260

Boutsika, K., Phillips, M. S., MacFarlane, S. A., Brown, D. J. F., Holeva, R. C., & Blok, V. C. (2004). Molecular diagnostics of some *trichodorid* nematodes and associated Tobacco rattle virus. *Plant Pathology*, *53*(1), 110–116. https://doi.org/10.1046/j.1365-3059.2003.00938.x

Boykin, L. M., Sseruwagi, P., Alicai, T., Ateka, E., Mohammed, I. U., Stanton, J.-A. L., Kayuki, C., Mark, D., Fute, T., & Erasto, J. (2019). Tree lab: Portable genomics for early detection of plant viruses and pests in Sub-Saharan Africa. *Genes*, *10*(9), 632.

Brito, J., Powers, T. O., Mullin, P. G., Inserra, R. N., & Dickson, D. W. (2004). Morphological and molecular characterization of *Meloidogyne mayaguensis* isolates from Florida. *Journal of Nematology*, *36*(3), 232.

Brown, T. A. (2002). Molecular phylogenetics. In *Genomes* (2nd ed.). Wiley-Liss. www.ncbi.nlm.nih.gov/books/NBK21122/

Charrier, E., Chen, R., Thundathil, N., & Gilleard, J. S. (2024). A set of nematode rRNA cistron databases and a primer assessment tool to enable more flexible and comprehensive metabarcoding. *Molecular Ecology Resources*, *24*(5). https://doi.org/10.1111/1755-0998.13965

Cold Spring Harbor Laboratory DNA Learning Center. (2020). *Using DNA barcodes to identify and classify living things*, October. https://dnabarcoding101.org/

Dababat, A. A., Paulitz, T., Laasli, S.-E., Lahlali, R., Li, H., Mokrini, F., & Dreisigacker, S. (2025). From genes to fields: Marker-assisted selection for nematode resistance in crops. *Integrative Plant Biotechnology*, *3*(1), 1–18. https://doi.org/10.55627/pbiotech.003.01.1129

De Ley, P., & Blaxter, M. (2002). Systematic position and phylogeny. In *The biology of nematodes* (pp. 1–30). CRC Press.

De Ley, P., De Ley, I. T., Morris, K., Abebe, E., Mundo-Ocampo, M., Yoder, M., Heras, J., Waumann, D., Rocha-Olivares, A., & Jay Burr, A. H. (2005). An integrated approach to fast and informative morphological vouchering of nematodes for applications in molecular barcoding. *Philosophical Transactions of the Royal Society B: Biological Sciences*, *360*(1462), 1945–1958.

Derycke, S., Remerie, T., Vierstraete, A., Backeljau, T., Vanfleteren, J., Vincx, M., & Moens, T. (2005). Mitochondrial DNA variation and cryptic speciation within the free-living marine nematode Pellioditis marina. *Marine Ecology Progress Series*, *300*, 91–103. https://doi.org/10.3354/meps300091

Dey, A., & Singh, S. (2009). Progress of science from microscopy to microarrays (Part 1): Diagnosis of parasitic diseases. *Journal of Laboratory Physicians*, *1*(1), 2–6.

Donn, S., Neilson, R., Griffiths, B. S., & Daniell, T. J. (2011). Greater coverage of the phylum Nematoda in SSU rDNA studies. *Biology and Fertility of Soils*, *47*(3), 333–339. https://doi.org/10.1007/s00374-010-0534-0

Elena, S. F., & Lalić, J. (2013). Plant RNA virus fitness predictability: Contribution of genetic and environmental factors. *Plant Pathology*, *62*(S1), 10–18. https://doi.org/10.1111/ppa.12102

Elmiligy, I. A., & Grisse, A. D. (1970). Effect of extraction technique and adding fixative to soil before storing on recovery of plant-parasitic nematodes. *Nematologica*, *16*(3), 353–358. https://doi.org/10.1163/187529270X00036

Esbenshade, P. R., & Triantaphyllou, A. C. (1985). Use of enzyme phenotypes for identification of Meloidogyne species. *Journal of Nematology*, *17*(1), 6.

Esbenshade, P. R., & Triantaphyllou, A. C. (1990). Isozyme phenotypes for the identification of Meloidogyne species. *Journal of Nematology*, *22*(1), 10.

Feng, X., Du, W., Luo, Q., & Liu, B.-F. (2009). Microfluidic chip: Next-generation platform for systems biology. *Analytica Chimica Acta*, *650*(1), 83–97. https://doi.org/10.1016/j.aca.2009.04.051

Floyd, R., Abebe, E., Papert, A., & Blaxter, M. (2002). Molecular barcodes for soil nematode identification. *Molecular Ecology*, *11*(4), 839–850.

Forensic and Scientific Services. (2007). *Method for Chelex extraction of DNA* (pp. 1–14). Forensic and Scientific Services.

Frézal, L., & Leblois, R. (2008). Four years of DNA barcoding: Current advances and prospects. *Infection, Genetics and Evolution*, *8*(5), 727–736. https://doi.org/10.1016/j.meegid.2008.05.005

Futai, K., Takeuchi, Y., & Kanzaki, N. (2005). A nested PCR-based method for detecting the pine wood nematode, Bursaphelenchus xylophilus, from pine wood. *Nematology*, *7*(5), 775–782. https://doi.org/10.1163/156854105775142928

Haegeman, A., Jones, J. T., & Danchin, E. G. J. (2011). Horizontal gene transfer in nematodes: A catalyst for plant parasitism? *Molecular Plant-Microbe Interactions*, *24*(8), 879–887.

Hajibabaei, M., Singer, G. A. C., Hebert, P. D. N., & Hickey, D. A. (2007). DNA barcoding: How it complements taxonomy, molecular phylogenetics and population genetics. *Trends in Genetics*, *23*(4), 167–172.

Hebert, P. D. N., Cywinska, A., Ball, S. L., & DeWaard, J. R. (2003). Biological identifications through DNA barcodes. *Proceedings of the Royal Society of London. Series B: Biological Sciences*, *270*(1512), 313–321.

Holterman, M., van der Wurff, A., van den Elsen, S., van Megen, H., Bongers, T., Holovachov, O., Bakker, J., & Helder, J. (2006). Phylum-wide analysis of SSU rDNA reveals deep phylogenetic relationships among nematodes and accelerated evolution toward crown clades. *Molecular Biology and Evolution*, *23*(9), 1792–1800.

Huang, D., Yan, G., Gudmestad, N., Ye, W., Whitworth, J., Frost, K., Crow, W., & Hajihassani, A. (2019). Developing a one-step multiplex PCR assay for rapid detection of four stubby-root nematode species, paratrichodorus allius, P. minor, P. porosus, and trichodorus obtusus. *Plant Disease*, *103*(3), 404–410. https://doi.org/10.1094/PDIS-06-18-0983-RE

Ijaz, S., & Khan, I. A. (2009). Molecular characterization of wheat germplasm using microsatellite markers. *Genetics and Molecular Research*, *8*(3), 809–815.

Ip, C. L. C., Loose, M., Tyson, J. R., de Cesare, M., Brown, B. L., Jain, M., Leggett, R. M., Eccles, D. A., Zalunin, V., & Urban, J. M. (2015). MinION analysis and reference consortium: Phase 1 data release and analysis. *F1000Research*, *4*, 1075.

Jabbar, A., Javed, N., Munir, A., Abbas, H., Khan, S. A., Moosa, A., Jabran, M., Adams, B. J., & Ali, M. A. (2021). Occurrence and molecular characterization of Meloidogyne graminicola on rice in Central Punjab, Pakistan. *Journal of Nematology*, *52*, e2020–e2123.

Jabran, M., Ali, M. A., Zahoor, A., Muhae-Ud-Din, G., Liu, T., Chen, W., & Gao, L. (2023). Intelligent reprogramming of wheat for enhancement of fungal and nematode disease resistance using advanced molecular techniques. *Frontiers in Plant Sciences*, *14*, 1132699. https://doi.org/10.3389/fpls.2023.1132699

Janssen, M., Van Der Voort, H., & Wahyudi, A. (2017). Factors influencing big data decision-making quality. *Journal of Business Research*, *70*, 338–345.

Kageyama, T., & Toju, H. (2022). Effects of source sample amount on biodiversity surveys of bacteria, fungi, and nematodes in soil ecosystems. *Frontiers in Ecology and Evolution*, 10. https://doi.org/10.3389/fevo.2022.959945

Kantor, C., Eisenback, J. D., & Kantor, M. (2024). Biosecurity risks to human food supply associated with plant-parasitic nematodes. *Frontiers in Plant Science*, *15*, 1404335.

Karssen, G., & Van Aelst, A. (2001). Root-knot nematode perineal pattern development: A reconsideration. *Nematology*, *3*(2), 95–111.

Kenmotsu, H., Uchida, K., Hirose, Y., & Eki, T. (2020). Taxonomic profiling of individual nematodes isolated from copse soils using deep amplicon sequencing of four distinct regions of the 18S ribosomal RNA gene. *PLoS One*, *15*(10), e0240336.

Kikuchi, T., Cotton, J. A., Dalzell, J. J., Hasegawa, K., Kanzaki, N., McVeigh, P., Takanashi, T., Tsai, I. J., Assefa, S. A., & Cock, P. J. A. (2011). Genomic insights into the origin of parasitism in the emerging plant pathogen Bursaphelenchus xylophilus. *PLoS Pathogens*, *7*(9), e1002219.

Kosaka, H., & Kanzaki, N. (2019). Insect parasitic nematodes, Contortylenchus sp. (Tylenchida: Allantonematidae) from the spruce bark beetle Ips typographus and the large larch bark beetle I. cembrae (Coleoptera: Scolytidae) in Japan. *Nematological Research (Japanese Journal of Nematology)*, *49*(2), 45–47. https://doi.org/10.3725/jjn.49.45

Kress, W. J., & Erickson, D. L. (2008). DNA barcodes: Genes, genomics, and bioinformatics. *Proceedings of the National Academy of Sciences*, *105*(8), 2761–2762. https://doi.org/10.1073/pnas.0800476105

Liu, J. U. N., Li, Q. I., Kong, L., Yu, H., & Zheng, X. (2011). Identifying the true oysters (Bivalvia: Ostreidae) with mitochondrial phylogeny and distance-based DNA barcoding. *Molecular Ecology Resources*, *11*(5), 820–830.

Liu, Z. (2007). Randomly amplified polymorphic DNA (RAPD). In *Aquaculture genome technologies* (pp. 21–28). Wiley. https://doi.org/10.1002/9780470277560.ch3

Lombard, L., Dekker-Nooren, C. C. E. M., Wuijster, B., van Kessel, S. P., van Duivenbode, I., van Bruggen, A. S., & van Heese, E. Y. J. (2024). Comparing the effectiveness of real-time PCRs to simultaneously detect and identify viable Globodera pallida and G. rostochiensis. *European Journal of Plant Pathology*, *169*(4), 681–696. https://doi.org/10.1007/s10658-024-02864-7

Long, E. O., & Dawid, I. B. (1980). Repeated genes in eukaryotes. *Annual Review of Biochemistry*, 49, 727–764, June. https://doi.org/10.1146/annurev.bi.49.070180.003455

McKay, A., Simpfendorfer, S., Vadakattu, G., Bithell, S., Moore, K., Daniel, R., Percy, C., White, J., Sparks, A., & Hollaway, G. (2018). *PREDICTA ® B update and new tests for 2018*. https://grdc.com.au/resources-and-publications/grdc-update-paper

Menegon, M., Cantaloni, C., Rodriguez-Prieto, A., Centomo, C., Abdelfattah, A., Rossato, M., Bernardi, M., Xumerle, L., Loader, S., & Delledonne, M. (2017). On site DNA barcoding by nanopore sequencing. *PLoS One*, *12*(10), e0184741.

Meyer, C. P., & Paulay, G. (2005). DNA barcoding: Error rates based on comprehensive sampling. *PLoS Biology*, *3*(12), e422. https://doi.org/10.1371/journal.pbio.0030422

Monis, P. T., Giglio, S., Keegan, A. R., & Thompson, R. C. A. (2005). Emerging technologies for the detection and genetic characterization of protozoan parasites. *Trends in Parasitology*, *21*(7), 340–346.

Mullis, K., Faloona, F., Scharf, S., Saiki, R., Horn, G., & Erlich, H. (1986). Specific enzymatic amplification of DNA in vitro: The polymerase chain reaction. *Cold Spring Harbor Symposia on Quantitative Biology*, *51*, 263–273.

Nadler, S. (2002). Species delimitation and nematode biodiversity: Phylogenies rule. *Nematology*, *4*(5), 615–625.

Navas, A., López, J. A., Espárrago, G., Camafeita, E., & Albar, J. P. (2002). Protein variability in Meloidogyne spp.(Nematoda: Meloidogynidae) revealed by two-dimensional gel electrophoresis and mass spectrometry. *Journal of Proteome Research*, *1*(5), 421–427.

Neher, D. A. (2001). Role of nematodes in soil health and their use as indicators. *Journal of Nematology, 33*(4), 161.

Niu, J., Guo, Q., Jian, H., Chen, C., Yang, D., Liu, Q., & Guo, Y. (2011). Rapid detection of Meloidogyne spp. by LAMP assay in soil and roots. *Crop Protection*, *30*(8), 1063–1069.

Odah, M. A. A. (2023). Unlocking the genetic code: Exploring the potential of DNA barcoding for biodiversity assessment. *AIMS Molecular Science*, *10*(4), 263–294. https://doi.org/10.3934/molsci.2023016

Palomares-Rius, J. E., Cantalapiedra-Navarrete, C., Archidona-Yuste, A., Blok, V. C., & Castillo, P. (2017). Mitochondrial genome diversity in dagger and needle nematodes (Nematoda: Longidoridae). *Scientific Reports*, *7*(1), 41813.

Pitcher, R. S., & Flegg, J. J. M. (1968). An improved final separation sieve for the extraction of plant parasitic nematodes from soil debris. *Nematologica*, *14*(1), 123–127.

Porazinska, D. L., Giblin-Davis, R. M., Faller, L., Farmerie, W., Kanzaki, N., Morris, K., Powers, T. O., Tucker, A. E., Sung, W. A. Y., & Thomas, W. K. (2009). Evaluating high-throughput sequencing as a method for metagenomic analysis of nematode diversity. *Molecular Ecology Resources*, *9*(6), 1439–1450.

Powers, T. (2004). Nematode molecular diagnostics: From bands to barcodes. *Annual Review of Phytopathology*, *42*(1), 367–383.
Powers, T. O., & Harris, T. S. (1993). A polymerase chain reaction method for identification of five major Meloidogyne species. *Journal of Nematology*, *25*(1), 1.
Ratnasingham, S., & Hebert, P. D. N. (2007). BOLD: The barcode of life data system. *Molecular Ecology Notes*, *7*(3), 355–364.
Roeber, F., Jex, A. R., & Gasser, R. B. (2013). Advances in the diagnosis of key gastrointestinal nematode infections of livestock, with an emphasis on small ruminants. *Biotechnology Advances*, *31*(8), 1135–1152.
Rubinoff, D. (2006). Utility of mitochondrial DNA barcodes in species conservation. *Conservation Biology*, *20*(4), 1026–1033. https://doi.org/10.1111/j.1523-1739.2006.00372.x
Saeki, Y., Kawano, E., Yamashita, C., Akao, S., & Nagatomo, Y. (2003). Detection of plant parasitic nematodes, Meloidogyne incognita and Pratylenchus coffeae by multiplex PCR using specific primers. *Soil Science and Plant Nutrition*, *49*(2), 291–295. https://doi.org/10.1080/00380768.2003.10410010
Savolainen, V., Cowan, R. S., Vogler, A. P., Roderick, G. K., & Lane, R. (2005). Towards writing the encyclopaedia of life: An introduction to DNA barcoding. *Philosophical Transactions of the Royal Society B: Biological Sciences*, *360*(1462), 1805–1811.
Seesao, Y., Gay, M., Merlin, S., Viscogliosi, E., Aliouat-Denis, C. M., & Audebert, C. (2017). A review of methods for nematode identification. *Journal of Microbiological Methods*, *138*, 37–49.
Shao, H., Zhang, P., Peng, D., Huang, W., Kong, L. A., Li, C., Liu, E., & Peng, H. (2023). Current advances in the identification of plant nematode diseases: From lab assays to in-field diagnostics. *Frontiers in Plant Science*, *14*, 1–12, January. https://doi.org/10.3389/fpls.2023.1106784
Sirangelo, T. M., & Calabrò, G. (2020). Next generation sequencing approach and impact on bioinformatics: applications in agri-food field. *Journal of Bioinformatics and Systems Biology*, *3*, 32–44.
Sonawala, D., Kapoor, A., Prabhu, A., Jain, U., Bhojwani, H., & Kumar, K. (2024). Improving cardiovascular outcomes in type 2 diabetes mellitus: Role of dapagliflozin and metformin. *International Journal of Cardiology*, *6*(2), 18–28.
Steinke, D., Prosser, S. W. J., & Hebert, P. D. N. (2016). *DNA barcoding of marine metazoans* (pp. 155–168). Springer. https://doi.org/10.1007/978-1-4939-3774-5_10
Subbotin, S. A., Roubtsova, T. V., Bostock, R. M., Tanha Maafi, Z., Chizhov, V. N., Palomares-Rius, J. E., & Castillo, P. (2023). DNA barcoding, phylogeny and phylogeography of the cyst nematode species of the Schachtii group from the genus Heterodera (Tylenchida: Heteroderidae). *Nematology*, *26*(1), 71–97. https://doi.org/10.1163/15685411-bja10292
Subbotin, S. A., Sturhan, D., Chizhov, V. N., Vovlas, N., & Baldwin, J. G. (2006). Phylogenetic analysis of Tylenchida Thorne, 1949 as inferred from D2 and D3 expansion fragments of the 28S rRNA gene sequences. *Nematology*, *8*(3), 455–474.
Thompson, C. H., McCartney, M. M., Roubtsova, T. V., Kasuga, T., Ebeler, S. E., Davis, C. E., & Bostock, R. M. (2021). Analysis of volatile profiles for tracking asymptomatic infections of phytophthora ramorum and other pathogens in rhododendron. *Phytopathology®*, *111*(10), 1818–1827. https://doi.org/10.1094/PHYTO-10-20-0472-R
Tigano, M. S., Carneiro, R. M. D. G., Jeyaprakash, A., Dickson, D. W., & Adams, B. J. (2005). Phylogeny of Meloidogyne spp. based on 18S rDNA and the intergenic region of mitochondrial DNA sequences. *Nematology*, *7*(6), 851–862.
Tileubayeva, Z., Avdeenko, A., Avdeenko, S., Stroiteleva, N., & Kondrashev, S. (2021). Plant-parasitic nematodes affecting vegetable crops in greenhouses. *Saudi Journal of Biological Sciences*, *28*(9), 5428–5433. https://doi.org/10.1016/j.sjbs.2021.05.075
Trasia, R. F. (2020). Nested polymerase chain reaction (Nested-PCR) as a diagnostic technique for intestinal helminth infection. *Jurnal Al-Azhar Indonesia Seri Sains Dan Teknologi*, *5*(4), 183. https://doi.org/10.36722/sst.v5i4.427

Treonis, A. M., Unangst, S. K., Kepler, R. M., Buyer, J. S., Cavigelli, M. A., Mirsky, S. B., & Maul, J. E. (2018). Characterization of soil nematode communities in three cropping systems through morphological and DNA metabarcoding approaches. *Scientific Reports*, *8*(1), 2004.

Vahidi, H., Curran, J., Nelson, D. W., Webster, J. M., McClure, M. A., & Honda, B. M. (1988). Unusual sequences, homologous to 5S RNA, in ribosomal DNA repeats of the nematode Meloidogyne arenaria. *Journal of Molecular Evolution*, *27*, 222–227.

Vancaester, E., & Blaxter, M. (2023). Phylogenomic analysis of Wolbachia genomes from the Darwin tree of life biodiversity genomics project. *PLoS Biology*, *21*(1), e3001972.

Van Den Hoogen, J., Geisen, S., Routh, D., Ferris, H., Traunspurger, W., Wardle, D. A., De Goede, R. G. M., Adams, B. J., Ahmad, W., & Andriuzzi, W. S. (2019). Soil nematode abundance and functional group composition at a global scale. *Nature*, *572*(7768), 194–198.

van de Vossenberg, B. T. L. H., van Kempen, T. L., van Duijnhoven, A. A. L. A. M., Warbroek, T., van der Gouw, L. P., Apolônio Silva De Oliveira, D., van Heese, E. Y. J., & Karssen, G. (2023). Development and validation of a high-throughput sequencing test for mitogenome and rDNA assembly and annotation, and its use in support of nematode identification of regulatory concern. *PhytoFrontiers*TM, 3(1), 103–112. https://doi.org/10.1094/PHYTOFR-04-22-0041-FI

Villar-Luna, E., Goméz-Rodriguez, O., Rojas-Martínez, R. I., & Zavaleta-Mejía, E. (2016). Presence of Meloidogyne enterolobii on Jalapeño pepper (Capsicum annuum L.) in Sinaloa, Mexico. *Helminthologia*, *53*(2), 155.

Vrain, T. C., Wakarchuk, D. A., Lévesque, A. C., & Hamilton, R. I. (1992). Intraspecific rDNA restriction fragment length polymorphism in the Xiphinema americanum group. *Fundamental and Applied Nematology*, *15*(6), 563–573.

Walsh, E. J., King, C., Grimes, R., & Gonzalez, A. (2005). Segmenting fluid effect on PCR reactions in microfluidic platforms. *Biomedical Microdevices*, *7*, 269–272.

Whyte, E. B., & Gowen, S. R. (1974). Other contributions: Recovery of nematodes from banana roots and soil samples. *Nematropica*, 27–31.

Wilson, E. O. (2003). The encyclopedia of life. *Trends in Ecology & Evolution*, *18*(2), 77–80.

Workentine, M. L., Chen, R., Zhu, S., Gavriliuc, S., Shaw, N., Rijke, J. de, Redman, E. M., Avramenko, R. W., Wit, J., Poissant, J., & Gilleard, J. S. (2020). A database for ITS2 sequences from nematodes. *BMC Genetics*, *21*(1), 74. https://doi.org/10.1186/s12863-020-00880-0

Yang, J., & Yu, G. (2019). A loop-mediated isothermal amplification assay for the plant-parasitic nematode Aphelenchoides besseyi in rice seedlings. *Journal of Nematology*, *51*(1), 1–11. https://doi.org/10.21307/jofnem-2019–080

Ye, W., Robbins, R. T., & Kirkpatrick, T. (2019). Molecular characterization of root-knot nematodes (Meloidogyne spp.) from Arkansas, USA. *Scientific Reports*, *9*(1), 15680.

Yuan, Z., Musa, N., Dybal, K., Back, M., Leybourne, D., & Yang, P. (2023). *Quantifying nematodes through images: Datasets, models, and baselines of deep learning*. 2023 IEEE 22nd International Conference on Trust, Security and Privacy in Computing and Communications (TrustCom), 2448–2455.

Zahoor, A., Niaz, Z., Tahir-ul-Qamar, M., & Hawamda, A. I. M. (2023). Computational analysis of 16D10 effector peptides from root-knot nematodes and their interaction with Mi resistance (R) protein through molecular docking. *Integrative Plant Biotechnology*, *1*(1), 1–8. https://doi.org/10.55627/pbiotech.001.01.0665

Zhang, C., Xing, D., & Li, Y. (2007). Micropumps, microvalves, and micromixers within PCR microfluidic chips: Advances and trends. *Biotechnology Advances*, *25*(5), 483–514. https://doi.org/10.1016/j.biotechadv.2007.05.003

Index

H

I

J

K

L

M

N

O

P

For Product Safety Concerns and Information please contact our
EU representative GPSR@taylorandfrancis.com Taylor & Francis
Verlag GmbH, Kaufingerstraße 24, 80331 München, Germany